REVISION IN OPTICS AND REFRACTION

Wai-Ching Leung MRCP(UK), DO, PGCE

Senior Registrar in Public Health,
Northern and Yorkshire Regional Health Authority,
Newcastle upon Tyne

ARNOLD

A member of the Hodder Headline Group
LONDON • SYDNEY • AUCKLAND
Co-published in the USA by Oxford University Press, New York

First published in Great Britain in 1997 by
Arnold, a member of the Hodder Headline Group,
338 Euston Road, London NW1 3BH

Co-published in the United States of America by
Oxford University Press, Inc.,
198 Madison Avenue, New York, NY 10016
Oxford is a registered trademark of Oxford University Press

British Library Cataloguing in Publication Data
A catalogue record for this book is available from the British Library

Library of Congress Cataloging-in-Publication Data
A catalog record for this book is available from the Library of Congress

ISBN 0 340 64613 6

Typeset in 10/12pt Helvetica by Gray Publishing, Tunbridge Wells
Printed and bound in Great Britain by J W Arrowsmith, Bristol

Contents

Preface		v
Introduction		1
Test 1	Questions 1–60	15
	Answers 1–60	27
Test 2	Questions 1–60	47
	Answers 1–60	59
Test 3	Questions 1–60	75
	Answers 1–60	87
Test 4	Questions 1–60	104
	Answers 1–60	116
Index		133

PREFACE

A few years ago, the aspiring ophthalmologist did not usually consider the knowledge and practice of optics and refraction essential acquisitions, as refraction was usually performed by opticians. In the past, this attitude was reflected in the content of the FRCS in Ophthalmology examinations. There was no special optics examination, and candidates often did not have to demonstrate knowledge and skills in optics and refraction in the general examination. While there was an optics and refraction section in the old Diploma of Ophthalmology, the knowledge and skills tested were relatively basic.

Since the establishment of the College of Ophthalmologists (now the Royal College of Ophthalmologists) in 1989, its Membership and Fellowship examinations have become regarded as necessary hurdles for every aspiring ophthalmologist or ophthalmic medical practitioner. Candidates are required to have passed the Optics and Refraction Module examinations before they are allowed to take the Membership examination or the final Fellowship examination. This Optics and Refraction Module spans almost a whole week, with a multiple choice question paper, a written short-answer-question paper, a viva voce examination, and a clinical examination in refraction. Candidates are expected to attempt this part of the examination quite early in their careers.

This emphasis on optics and refraction for ophthalmologists is overdue. Refractive or accommodative abnormalities are probably the most common cause of visual complaints. In order to determine the causes of visual symptoms, the ophthalmologist must exclude refractive errors or accommodative abnormalities. Furthermore, patients rightly expect the ophthalmologist to possess sufficient skills for prescribing the appropriate treatments for their eye problems.

Unfortunately, the importance of optics and refraction is not always reflected in the training given to the trainee ophthalmologist. There are many reasons for this. First, demonstration of the skills involved is quite time-consuming and the workload of most ophthalmology outpatient clinics often makes this impractical. Second, frequent practice under supervision is vital for acquiring the necessary skills. Third, while consultant ophthalmologists usually run teaching sessions on clinical ophthalmology, it is rare for them to organise tutorials on optics and refraction. Hence, trainees generally have to pick up both the theoretical knowledge of optics and the practical skills of refraction on their own.

While there are a few textbooks available on the theory of optics, there is at present no revision book for this important examination module. Candidates therefore have no opportunities to practise multiple choice questions before their examination, and they have little guidance on the practical component. The practical examination is especially important, as failure on this would not be compensated by outstanding performance on other parts of the examination.

This book attempts to address some of these problems, and is split broadly into two sections. The first section explains the format of the examination, and then provides detailed guidance on preparing for the practical examination and the techniques of refraction. Although this will act as a guide, there is no substitute for frequent practice under supervision. The second section consists of four multiple choice tests of 60 Questions each, followed by Answers and Explanations. The candidates should answer 'true', 'false' or 'don't know' to each of the five stems of each question. Detailed explanations are given so that the may be used as revision for the viva voce examination and short-answer questions.

I hope candidates approaching this important examination will find this book of use.

Wai-Ching Leung

INTRODUCTION

The Royal College of Ophthalmologists Optics examination

The examination takes place twice a year, over a period of 4 days. On the first day, there is an essay paper in the morning, and a multiple choice question paper in the afternoon. On the second day, there is an oral examination. The practical refraction examination may take place either on the third or the fourth day, and the results are announced at the end of the fourth day.

Written paper

Two hours are allowed for this paper. There are four main questions in the paper, all of which are compulsory. There is usually one question on an optical instrument, one or two questions on practical refraction and prescription, and one question with an emphasis on basic science in optics. One or two of the questions may be subdivided into two or three parts. It is important to read the questions carefully, and only answer what is asked. When you are asked to discuss the causes of a topic, it is advisable to have a paragraph for each cause, and underline each cause to make it easier for the examiner. You will be asked to draw simple ray diagrams in some of the questions.

Multiple choice questions

There are 60 questions to be answered in 2 hours, similar to the tests in this book. About half of the questions are on optics, and half on practical refraction.

Oral examination

There are two examiners, both of whom are clinical consultant ophthalmologists. Both optics and practical refraction questions may be asked, with particular emphasis on the latter. This part of the examination lasts for 20 minutes, with each examiner questioning for 10 minutes.

Practical refraction

You will spend 15 minutes with the patient on your own, and the remaining 15 minutes will be in the company of the examiners, who are clinical consultant ophthalmologists. This is discussed below.

Clinical Examination in Refraction

Introduction

The optics and refraction examination is one module of both the Membership and Fellowship examinations by the Royal College of Ophthalmologists, as well as part of the Diploma of Ophthalmology examination in Dublin. The requirements for the Royal College of Ophthalmologist examination are possibly more stringent. The candidate is expected, in about 30 minutes, to take a very brief relevant history and perform the necessary examination of a subject (patient) in order to decide the optimum optical correction. For the Royal College of Ophthalmologists examination, the candidate spends 15 minutes alone with the patient. The remaining 15 minutes will be in the company of the examiners. The examiners will check the candidate's results and refraction techniques, especially the objective refraction with a retinoscope and the clear lucid questioning of the patient in the subjective refraction. The examiner may then ask questions relating to the history, refraction technique, results obtained, and management of the subject.

As a full history and refractive examination would undoubtedly take longer, the candidate has to prioritise various points in the history and examination. For example, the examiner is unlikely to be impressed by a candidate who did a thorough ophthalmic history and performed a full muscle balance examination but did not perform retinoscopy through lack of time. Therefore, it is important for the candidate to have a clear perspective as to what is vital and what is less important in the examination. The candidates should decide well in advance of the examination in what order they will tackle the various activities, and to practise them over and over again. As a general advice, it is much better to be fluent in one particular way of taking the history and performing the refractive technique than trying out a variety of ways less fluently. The candidates should be thoroughly familiar with one way of recording results. For example, the candidate may decide in advance whether the retinoscopy results would be recorded in the format of an astigmatic cross diagram or in the shorthand form of, say, 3DS/2DC × 30°, or as a 'fraction'.

On the other hand, the candidates must be prepared for refraction of all age groups, from a young child of, say, 8 years old to an elderly person. The requirements and techniques amongst these age groups differ slightly. Be prepared for situations when retinsocopy is difficult (e.g. when there are lens opacities), or when the subject is not as cooperative as one might wish.

Although nearly all examination centres provide the full set of instruments, it saves time in the examination to provide your own reliable equipment with which you are familiar. This applies especially for the retinoscope, ophthalmoscope and trial frame. Check that your instruments are in good working order, and that the batteries for the retinoscope and ophthalmoscope are fully charged the day before the examination. A checklist for the equipment required is provided.

It is the purpose of this section to suggest one way of tackling these problems. Candidates, however, may well have their own personal methods. This is fine as long as the methods are well thought out and well practised.

Preparing for the clinical examination

The clinical examination is a test for **practical** skills. The candidates must demonstrate their ability to apply their skill fluently and efficiently, their ability to work in limited time, their ability to adapt in situations which are not theoretically ideal (e.g. with uncooperative subjects, those with opacities in the media, etc.), and the use of common sense and judgement.

These skills cannot be acquired from reading the most detailed textbooks. They must be acquired from practice. Some trainees have adequate opportunities for performing refraction under their consultants' guidance. It is very useful to receive feedback about their techniques and refraction results. However, others are less fortunate and have little such opportunities. This means that even if the trainees are keen to practise on clinical refraction, they would not have any feedback as to whether their results or techniques are correct or not.

These are real problems, but not insurmountable. Possibilities for overcoming them are include the following.

1 Arrange with the hospital opticians to sit in their clinics, and then refract under their supervision. The usual stumbling block is that the timetable may not allow this. Try to ask your consultants for a few days study leave for this purpose, or to have one afternoon a week for this purpose.
2 Ask your registrar or senior registrar to give you a few sessions on clinical refraction. Again, the problems may be that both they and you may be too busy.
3 Go on courses for optics and refraction. However, it may be fair to say that although many of the courses are extremely good in teaching the theory, there may not be enough opportunities for practice. This is understandable in view of the short length of time available, and that it would be difficult to arrange for a large number of willing subjects to be refracted.
4 If all the above options fail, one can still acquire the skills by determining to refract at least 2–3 subjects a week. Start with the 'easy subjects' first—those who are cooperative, with no opacities in the media, and with moderate refractive errors only. They may be in-patients or out-patients. For in-patients, patients with post-operative cataract extraction are most suitable to begin with: their optical media are usually clear, they are generally available for a reasonable period of time, and you can generally tell roughly whether you have got the right retinoscopy results by the visual acuity obtained. Also, these patients are often used for the examination. The disadvantages are that you cannot compare your results with those obtained by an optician or your consultant, and that sometimes there may be very significant cylindrical correction, which may be difficult for the beginner. For out-patients, the best subjects are those you are referring to the opticians, including those for post-operative cataract extraction. Try to arrange for them to see you at the end of the clinic, so that you can spend a bit longer time with them. Note down your refractive results, and compare them with the results obtained by the opticians. If there are significant difference, try to refract once more and discover the reasons for the discrepancies.

Initially, you can concentrate on the retinoscopy findings only. When you have acquired a reasonable skill in retinoscopy, then gradually proceed to both subjective and objective distant refraction. This will be less tiring for you and your patients, and you can concentrate on one task at a time. When you are confident with these, proceed with near correction and measurement of amplitude of accommodation.

As you gain confidence in the 'easy subjects', and generally obtain the correct refractive results, then gradually look for 'less easy' subjects, such as children, those with high myopia or hypermetropia, those with high cylindrical corrections, those who are less cooperative, and those with opacities in the media. It is probably best to tackle one different problem (e.g. high myopia) at a time, to gain the techniques of refraction in this group of subjects.

When you are generally confident about these, develop your own routine (one is suggested in this section), so that you are fluent and efficient in this. Decide on your working distance and the position of the mirror in your retinoscope, and adhere to them. Develop your own routine of recording your findings (one suggestion is given in this section). Keep practising them, and start timing yourself to make sure that you can complete the refraction in the time allotted in the examination.

Once you can do this, you are ready for the examination.

Checklist of equipment for use in the clinical refractive examination

1 Retinoscope (battery fully charged)—a streak retinoscope is probably the easiest to use.
2 Direct ophthalmoscope (battery fully charged).
3 Pen torch.
4 (a) Eye occluder for one eye and pinholes for the other.
  (b) Fixation target.
5 Tape measure (for measuring near point).
6 Near vision acuity chart.
7 Cross-cylinders.
8 Pen and paper.

The practical examination

It is important to dress smartly and conservatively in the practical examination.

The examiners will introduce you to your patient. Be courteous, introduce yourself and explain that you are taking an examination. Explain that you have limited time to complete your tasks, and therefore you may sometimes need to be selective in what you want to ask and do. Figure I.1 shows a suggested recording sheet which serves as a useful checklist.

Suggested recording sheet

History:

Name: Age:

Occupation: Hobbies:

(Brief previous ophthalmic operations):

Examination

	Right	**Left**
VA without glasses		
VA with glasses (if available)		
VA with pinhole		

Objective method

Retinoscopy—lens used	___DS / ___DC × ___deg	____DS / ___DC × __deg
Working distance	________ metres (equivalent to ____D)	_________ metres (equivalent to ____D)
Net retinoscopy result	____DS/ ___DC × ___deg	___DS / ___DC × __deg

Subjective methods

Final results	___DS/___DC × ___deg	____DS/___DC × ___deg
Corrected visual acuity	____________	_____________
Binocular visual acuity	________________	

Near correction

(Amplitude of accommodation)	_______D	________D
Near correction given	+______D	+_______D
Reading visual acuity (with corr)	_________	__________
Binocular reading visual acuity	________________	

Figure I.1 Suggested recording sheet (continued overleaf).

Other activities

Muscle balance	**Right**	**Left**
Cover test (distant)		
(near)		
Alternative cover test		
Maddox Rod (or prism cover test)		

Fusion

Worth's 4 dot test

Binocular balance

Dissociation with prisms
+Duochrome test

Fundoscopy

Figure I.1 Suggested recording sheet.

Taking history

As this is primarily an examination of refractive skills, the history should be brief, and proceed with the examination as early as possible. Ask for the following information.

Name
Age
Occupation and hobbies
Previous ophthalmic operations—especially cataract operations.
Refractive errors—you can always try to ask whether they are long-sighted, short-sighted or astigmatic. Alternatively, you can ask whether they need glasses for distance only, for reading only, or all the time. Sometimes, patients have been instructed not to tell the candidates.

Refractive examination

Visual acuity

Check the distant visual acuity of each eye unaided and with glasses, if available. Record the visual acuity in standard form (e.g. 6/36, 6/9, CF at 1 m, etc.). Sometimes, patients are told not to show candidates their spectacles. However, if it is available, quick neutralisation of each lens would give a rough idea of the refractive errors. This is especially true if the visual acuity obtained with spectacles is much better than that unaided.

You should have a rough idea of the degree of the refractive errors by unaided visual acuity alone. If the unaided acuity is good, the patient cannot have a high refractive error.

If the unaided acuity is poor, you can perform a pinhole acuity. With minor exceptions, an improvement would indicate that the visual acuity can be corrected by refraction. On the other hand, you may decide to proceed with the refraction, and measure the pinhole acuity only if there are no improvements in visual acuity with refraction.

Objective refraction—retinoscopy

Position the trial frame so that it is comfortable for the patient, and is reasonably well centred. There is insufficient time to use a proper centring device. From the information gleaned from neutralisation of the subject's present spectacle lenses or pinhole improvement of visual acuity, you may already have some ideas of the approximate correction required for the patient.

Occlude one eye, say the left eye. Darken the room, and ensure that both you and the patient are sitting comfortably. Next, you should encourage your patient to relax the accommodation. Ask the patient to fixate at a spot of light at a distance. Alternatively, you may decide to use the fogging technique by putting a highly convex lens in front of the left eye.

You should be familiar with your working distance, and know its equivalence in dioptric power. (For example, a working distance of 0.5 m is equivalent to +2D; a working distance of 2/3 m is equivalent to +1.5D.) Put the appropriate corrections (estimated refractive error + working distance equivalence) in the trial frame. For example, if the estimated refractive error is +2D (hypermetropia) and the working distance is 2/3 m, you should put in (2 + 1.5) = +3.5D in the trial frame.

Ensure that the mirror of your retinoscope is in the position familiar to you. Refract the right eye. Find a meridian in which the band outside the pupil is in alignment with the band in the pupil. (If there is no astigmatism, this occurs in all meridians. If there is a high degree of astigmatism, it only occurs in two meridians perpendicular to each other.) Find the point of neutralisation. Remember that you are nearer to the neutralisation point if the reflex is brighter and larger, and moves faster. Once you neutralise one meridian, turn the beam 90° and neutralise the other meridian by adding cylinders with the axis parallel to the beam. Then record the lens used and the axis of the cylinder in the format _____ DS/_____DC @ _____ degrees. You may prefer to use the 'cross' or 'fractional' notation. You may use them as long as you are familiar and consistent with their use.

Next, perform the same procedure with the left eye. Again record the lenses used.

Then, calculate the net correction required by subtracting the appropriate dioptric power for the working distance. For example, if lenses –4DS/–1DC@20° were used, and the working distance is 2/3 m, deduct 1.5D from the spherical power. Hence, the net correction required is –5.5DS/–1DC@20°.

Problems with retinoscopy

1 No retinoscopy reflex seen at all

You can try the following:

a. perform fundoscopy with the direct ophthalmoscope both to look for causes (e.g. lens opacity) and to estimate the refractive error roughly by the power of the lens you need to put up to visualise the fundus. Repeat retinoscopy using the approximate correction obtained.
b. Quickly try to observe for retinsocopy reflex with lenses of power –5D, –10D, –15D, +5D, +10D, +15D. Once a reflex is seen, you can then change the correction lenses by smaller steps.

2 Small pupils and interference from other reflexes

Darken the room more, and try to alter the working distance or tilt of the beam slightly.

3 Difference in neutralisation point at the centre and the periphery of the pupil

Try to concentrate on the centre of the pupil, and ignore the appearance at the periphery.

4 Scissors movement

Irregular astigmatism due to corneal irregularity may be the reason. Concentrate on the central part of the pupil.

Subjective refraction

1 Check the visual acuity corrected with the net correction required, as calculated above. Use of subjective refraction technique to refine the correction.

Spherical component

2 If the visual acuity is 6/12 or better, alter the spheres by steps of about 0.5D each time. Find the **most convex** or **least concave** correction which would give the best visual acuity in the right eye while the left is occluded. Then repeat the procedure with the left eye with the right occluded. It is best to postpone the Duochrome test until after the cylindrical correction is verified.

Cylindrical component

Axis

3 If the subject is cooperative and perceptive, the best method is verifying the cylindrical component by cross-cylinder. To verify the **axis** of the cylinder, ask the subject to look at the letter 'O' on the Snellen chart. Place the cross-cylinder, with the **handle** of the cross-cylinder, in alignment with the axis of the trial cylinder. Then flip the handle of the cross-cylinder 180° and compare the visual effect between the two positions (see Fig. I.2).

If the visual effect is better in one position, hold the cross-cylinder in this position. Rotate the trial cylinder in the direction marked on the cross-cylinder

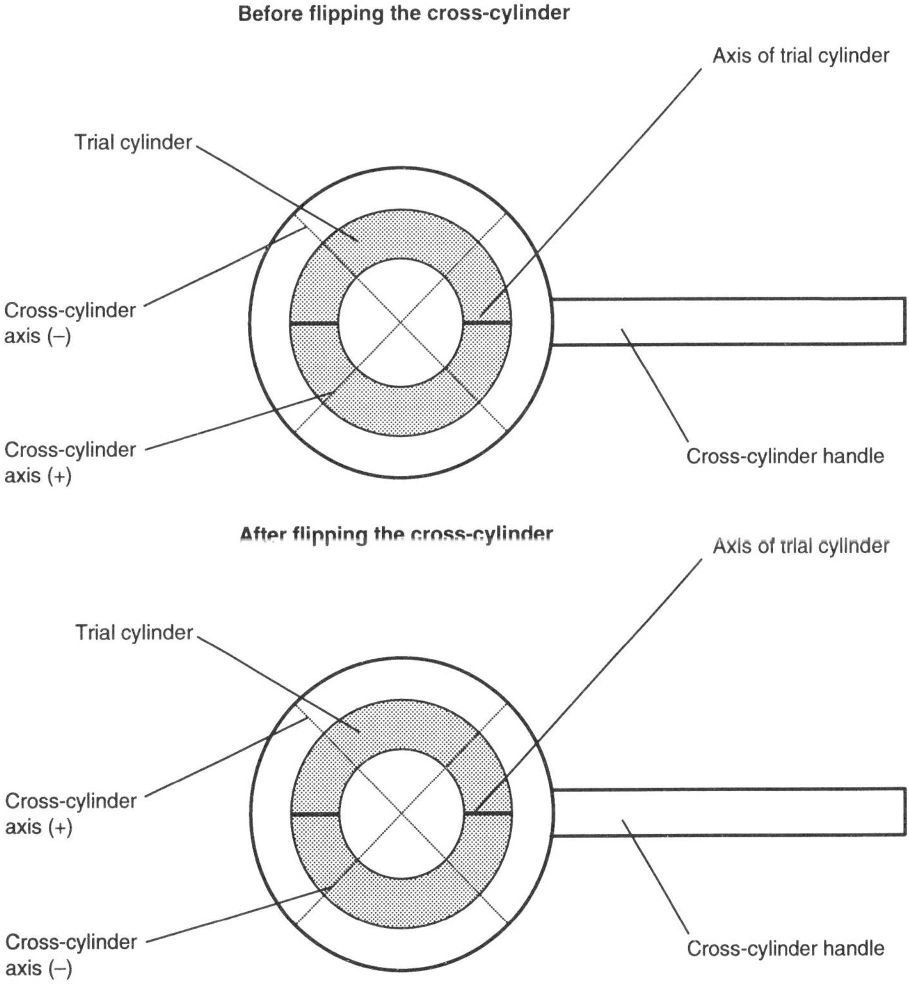

Figure I.2 Use of the cross-cylinder in verifying axis.

with the **same sign** as the trial cylinder. For example, assume you are using the **plus** cylinder in the trial frame. Hold the cross-cylinder in the position with the better visual effect. Turn the trial cylinder in the direction marked **plus** on the cross-cylinder. If the power of the cylinder used is high, rotate in steps of no more than 5–10°. If the power used is low, you may need to rotate through a larger angle for the subject to detect any effect. The process is repeated until there is no difference in the visual effects when the handle of the cross-cylinder is flipped.

Power

4 Verify the **power** of the trial cylinder by placing the cross-cylinder with the axis of the trial cylinder in alignment with the axis of the cross-cylinder. Again, ask the patient to look at the letter 'O' on the Snellen chart, and the cross-cylinder is flipped 180°. The subject compares the visual effects. If the subject prefers one position to the other, hold the cross-cylinder in this position. Change the power of the trial cylinder in the **same sign** as the sign of the axis of the cross-cylinder in alignment with the trial cylinder. For example, assume that the visual effect is better when the **negative** axis of the cross-cylinder is in alignment with the axis of the trial cylinder. You should change the power of the trial cylinder by making it more **negative.** That is, reduce the power of the cylinder if a positive trial cylinder is used, or increase the power of the cylinder if a negative trial cylinder is used. Do not change the power more than 0.5D at a time (see Fig. I.3).

It is important that when the power of the cylinder is changed, the spherical lens of opposite sign and half the power of the cylinder is added. Otherwise, the posterior focal plane would move further away from the eye, and the image becomes blurred.

The process is repeated until there is no difference in the visual effects between the two positions when the cross-cylinder is flipped.

Recheck axis of cylinder

5 Recheck the axis of the cylinder if the power is changed.

Recheck spherical component

6 Recheck the spherical component by the best vision sphere method as in (2). Then check with the Duochrome test. First, encourage the subject to relax the accommodation either by darkening the room and fixating at a distant light, or by placing an extra convex lens (of about 2D) with the corrections in front of both eyes for a minute (the fogging method). Secondly, with the correction lenses, ask the patient to look at the two rows of black letters illuminated on green and red backgrounds, and decide which row of letters is clearer and darker. The first impression is most important, before the subject has a chance to accommodate. If the red row of letters is darker and clearer, reduce the negative sphere (or increase the positive sphere) by no more than 0.25D at

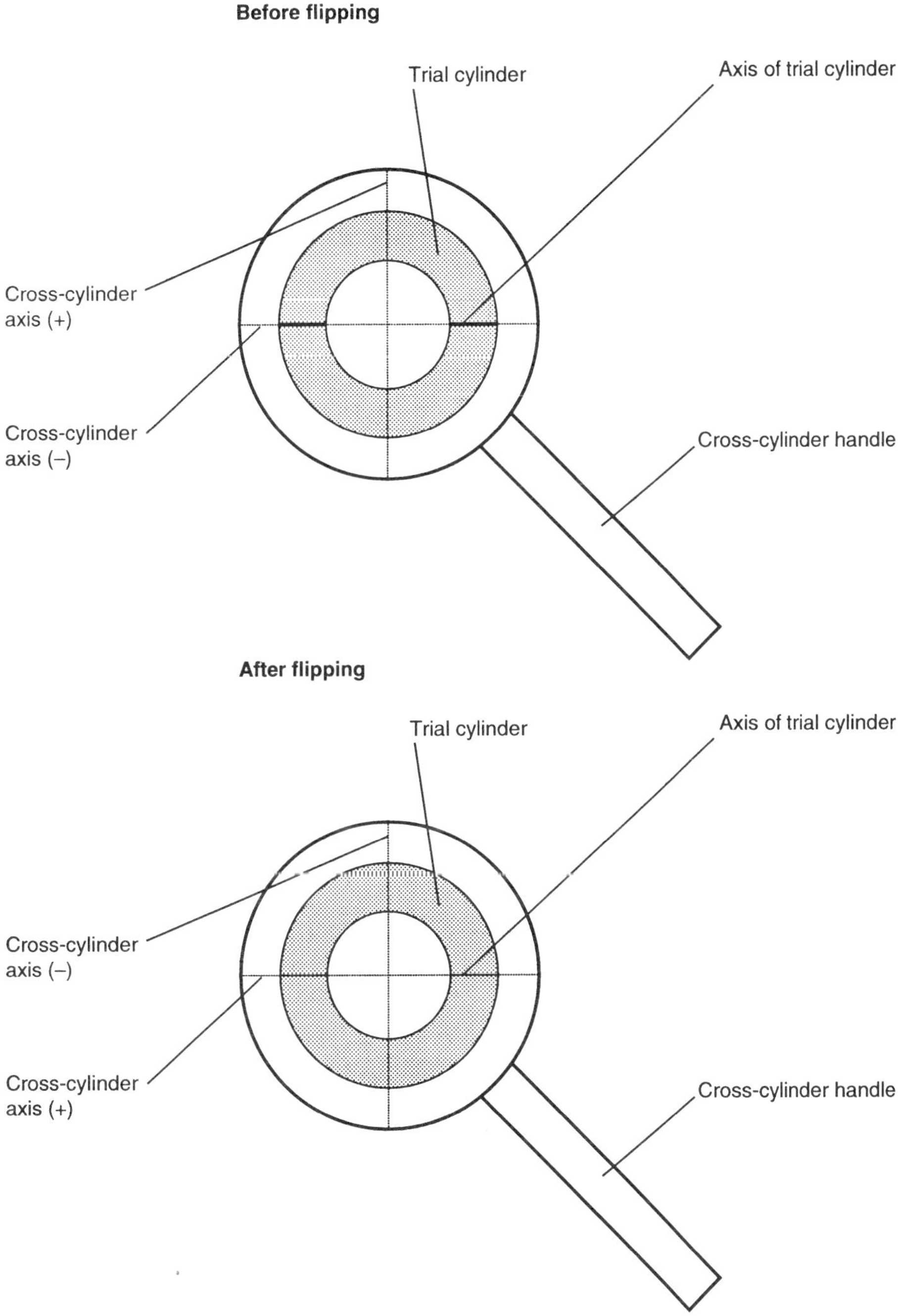

Figure I.3 Verifying the power of the cross-cylinder.

a time. Conversely, if the green row of letters is darker and clearer, increase the negative sphere (or reduce the positive sphere) until the two rows of letters are equally dark.

Repeat the procedures with the other eye.

7 Record the lens used after subjective refraction, and the visual acuity achieved.
8 Record the binocular visual acuity achieved.

Problems you may encounter

a. Patient cannot tell the visual difference between the two positions on flipping the cross-cylinder

This may be because the astigmatism is entirely corrected. This can be verified by changing the trial cylinder, and the patient should then notice a difference. However, it may also be due to the patient's low power of perception, or because the visual acuity achieved so far is poor. If this is the case, use a 'stronger' cross-cylinder (i.e. +0.5DS/–1.0DC rather than +0.25DS/–0.5DC).

If this fails, you can try the astigmatic fan. Leave the patient over-corrected by about +0.5D sphere, and ask him/her to look at the fan and decide which line or group of lines are clearer and darker than the others. You can then neutralise this with negative cylinder at 90° to this line.

b. Patient cannot relax accommodation in the best vision and Duochrome test

Try to fog the subject with an extra convex lens (say, of about 2–3D) until the visual acuity is about 6/60. Then gradually reduce the convex lens until the best visual acuity is achieved.

Near correction

This should be considered in any patients over the age of, say, 30. The general principles are as follows.

a. General comfort for the patient, the occupation and hobbies of the patient, are more important than theoretical considerations.
b. It is better to under-correct than over-correct for near correction, especially for those who do not have reading spectacles previously.
c. The correction should be equal in both eyes. Corrections exceeding 3.5D are seldom necessary.

General procedures for near correction

It is unlikely that you would have sufficient time to perform the full examination of amplitude of accommodation and convergence. The following simplified procedures are sufficient.

a. If the near visual acuity with a distance correction is good with a reasonable near point, no near correction is needed.
b. You can rapidly assess the amplitude of accommodation by assessing the

near point by bringing the reading text closer to the patient with distant correction until blurring occurs. For example, if the near point is 33 cm, the amplitude of accommodation is about (1/0.33) = 3D.

c. Calculate the approximate amplitude of accommodation required, taking into account the patient's hobbies and occupation. For example, if the desired reading distance is 33 cm, the required amplitude of accommodation is (1/0.33) = 3D
d. Aim to keep about 1/3 of the available accommodation in reserve. In example b. above, only (2/3 × 3) = 2D of the available accommodation should be used.
e. Calculate the near correction needed from the difference of c. and d. above. In the above example, the near correction required is (3 – 2)D = +1D.
f. Try the calculated near correction, and measure the near visual acuity. Test for comfort. Aim for undercorrection rather than overcorrection.

Other activities

Cover test

The cover test is performed with the patient viewing a distant target and then repeating with a near target. The subject is asked to fixate on an object, and look for movement of the left eye when the right eye is being covered. Look for movement of the right eye when it is being uncovered. This will reveal a manifest squint.

Alternate cover test

This detects the presence of a latent squint. One eye is covered for 2 s, and the cover is quickly shifted to the other eye. Look for a movement of the **uncovered** eye.

Maddox rod

The Maddox rod is placed in front of the right eye to dissociate the eyes, while the subject views a white light 6 m away. The image from the right eye appears as a red line. If the subject reports the red line to the left of the white point, the subject has a divergent squint. If the subject reports the red line to the right of the white point, the subject has a convergent squint. Prisms are added until the red line and the white point coincide, and the degree of the squint can be measured.

Prism cover test

This can be used instead of the Maddox rod, and is more rapid. Neutralise the eye movements in the cover test with a prism of suitable power, and the power of the prism is a measure of the squint.

Worth's four dot test for fusion

The subject wears red-green goggles with red in front of the right eye and green in front of the left eye. The subject views a box with four lights (one red, two green, one white).

Report	*Interpretation*
four lights, no manifest squint	Normal fusion
four lights, manifest squint	Abnormal retinal correspondence
two red lights only	Left eye suppression
three green lights only	Right eye suppression
five lights (three green, two red)	Diplopia

Binocular balance

A simple binocular balance test is to use a prism to dissociate the eyes, and use the Duochrome test. The subject is asked whether the two rows of letters with the green background illumination and two rows of letters with the red background illumination are equally dark. The spheres can then be adjusted. This can be combined with the fogging technique (adding an extra +2D lens, and gradually reducing it by steps of 0.5D, until the maximum visual acuity is obtained).

TEST 1
Questions

1 Regarding the visible spectrum of light

A the wavelength is roughly between 400 and 700 nm
B red light has a shorter wavelength than blue light
C a greater proportion of light with a wavelength below 400 nm reaches the retina in the aphakic eye compared with the phakic eye
D the pigment epithelium of the iris absorbs light radiation of most wavelengths
E sunlight contains light of different wavelengths

2 The refractive index of a medium is

A the ratio of sin *r* to sin *i*, where *i* is the angle of incidence from air to medium and *r* is the refracted angle
B the ratio of velocity of light in air to the velocity of light in the medium
C a measurement of optical density of the medium
D determinable from the critical angle from medium to air
E higher in crown glass than water

3 Diffraction

A is a property of particles
B increases when a wavefront is projected through a large opening
C refers to the interference of secondary waves formed at the edge of the aperture with the main wavefront
D does not occur through a small lens
E can limit the visual acuity obtained by a pinhole

4 When a ray of light enters a parallel glass block

A the emergent ray will be slightly less in intensity than the incident ray
B the emergent ray is parallel to the incident ray
C the displacement of the emergent ray increases as the thickness of the glass block increases
D the displacement of the emergent ray increases as the incident angle decreases
E total internal reflection may occur on the other side of the glass block

5 In the reduced eye

A there are two nodal points
B the first focal point is about 24 mm in front of the cornea
C the second focal point is about 24 mm behind the cornea
D the nodal point(s) lie between the principal point(s) and the retina
E there is only one principal point

6 The following lenses are equivalent

A	+3DS/+2DC at 60°	and	+5DS/–2DC at 60°
B	–3DS/+2.5DC at 140°	and	–0.5DS/–2.5DC at 50°
C	–4DS/+2DC at 30°	and	–2DS/+2DC at 150°
D	–3DS/–1DC at 80°	and	–4DS/+1DC at 170°
E	+2DS/–1DC at 90°	and	+2DS/+1DC at 0°

7 The effect of addition of a cross-cylinder to the conoid of Sturm may include

A increasing the interval of Sturm
B decreasing the interval of Sturm
C moving the circle of least diffusion away from the lens
D moving the circle of least diffusion towards the lens
E moving the circle of least diffusion outside the interval of Sturm

8 The accommodative–convergence:accommodation ratio

A varies from one minute to the next for the same individual
B is different for different individuals
C cannot be measured clinically
D is on average about 3.5 to 4 Δ of convergence for 1D of accommodation
E should be taken into account in deciding the near correction for presbyopia

9 In identification of the type of lenses by comparing the visual image of a cross viewed through the lens

A a convex lens causes distortion of the cross
B a convex lens causes a 'with movement'
C a concave lens causes an 'against movement'
D an astigmatic lens causes distortion of the cross
E a rotation of a strong convex lens causes scissoring of the cross

10 Which of the following statements is/are true about astigmatism?

A The principal meridians in regular astigmatism may be at 45° to each other.
B If the axes of the principal meridians of each eye are the same, the astigmatism is symmetrical.
C Asymmetrical astigmatism is more common than symmetrical astigmatism.
D Uncorrected asymmetrical astigmatism is more likely than uncorrected symmetrical astigmatism to cause torticollis.
E Irregular astigmatism can be totally corrected by cylinders.

11 An aphakic patient requires a contact lens of +14D.

A A spectacle lens of about 11.5D is required if the back vertex distance is 15 mm

B A spectacle lens of about 17.75D is required if the back vertex distance is 15 mm.

C A spectacle lens of 14D is required irrespective of the back vertex distance

D A spectacle lens of about 12D is required if the back vertex distance is 12 mm.

E A spectacle lens of about 16.25D is required if the back vertex distance is 10 mm.

12 The far point

A is the position of an object such that its image falls on the retina of an eye in the absence of accommodation

B of a hypermetropic eye is real

C of a myopic eye is a finite distance in front of the eye

D of a –2D myopic eye is 50 cm in front of the eye

E of an emmetropic eye is at infinity

13 Myopia

A is of the axial type in keratoconus

B may be due to the eye being abnormally short

C is associated with convergence of parallel rays entering the cornea at a point behind the retina

D can be corrected by a convex lens

E requires a stronger lens for correction the further away the lens is from the eye

14 The following is/are true regarding hypermetropia.

A Total hypermetropia is the total of facultative and absolute hypermetropia.

B Latent hypermetropia is the difference between total and manifest hypermetropia.

C Latent hypermetropia is the difference in refraction of the total relaxed eye with and without atropine.

D It is mostly facultative in an infant.

E Latent hypermetropia is related to the tone of the ciliary muscle.

15 Astigmatism

A requires optical correction if it is an asymptomatic simple type of 0.25D

B requires correction if it causes 'eye strain'

C should be corrected for distant vision only

D causes more reduction in visual acuity in compound hypermetropic type than mixed astigmatism

E can be corrected by contact lenses

16 Accommodation

A is associated with dilatation of the pupil
B occurs equally in both eyes
C is associated with convergence of the eyes
D may result in a convergent squint
E occurs simultaneously in both eyes

17 In pathological axial myopia

A myopia appears from childhood
B it may progress to more than 20D
C degenerative changes of fundus characteristically become widespread in adolescence
D it is more common in Japanese
E it may be genetic in origin

18 The following is/are true regarding accommodation.

A It increases as age increases.
B It is mainly brought about by changes in curvature of the posterior surface of the lens.
C The range of accommodation remains static in the first decade of life.
D It is reduced by the use of cyclopentolate eye drops.
E It is associated with slackening of the zonules.

19 The following is/are true regarding the near point.

A It is the closest point to the eye at which small objects can be seen without blurring.
B Objects closer to the eye than the near point can be clearly seen by accommodation.
C It is further away in the myopic eye than the emmetropic eye.
D It is nearer in the emmetropic eye than the hypermetropic eye.
E It moves further away as age increases.

20 A Fresnel prism

A consists of two prisms
B is about 2 mm thick
C is used for inverting images
D is used for assessing microtopia
E is usually made of crown glass

21 Total internal reflection can be used to explain

A how light travels in fibre optics
B the formation of a secondary rainbow
C the use of a triangular prism as a mirror
D the formation of an image from a convex mirror
E antireflection coatings of lenses

22 Difficulties of correction of aphakia by spectacles include

A prismatic errors from displaced optical centres
B spherical aberration
C image diminution
D incorrect prescription due to inaccurate estimate of vertex distance
E reduced visual fields

23 Back vertex distance

A is the distance between the front and the back of a spectacle lens
B is the distance between the plano side of a spectacle lens and the cornea
C is especially important to be measured accurately in aphakia
D should be measured for myopia of 1.5D
E is the same for every subject

24 Clinical measurement of visual acuity

A measures the threshold of discrimination of two spatially separated targets
B assesses the function of the rods
C can also be expressed by the angular threshold of discrimination of two spatially separated targets subtended at the nodal point of the eye
D by Landolt's rings is heavily influenced by educational background
E by Snellen's chart is less confusing to the subject than Landolt's rings

25 The cross-cylinder

A is a type of toric lens
B the same as superimposing two cylindrical lenses of equal power but opposite sign with their axes at 45°
C can be used to check the axis of the cylinder by placing the handle in line with the axis of the trial cylinder, and comparing the visual effect when it is flipped over
D can be used to check the power of the cylinder by placing the handle in line with the axis of the trial cylinder, and flipping over to compare the visual effect
E can only be of the type 0.5DS/–1.0DC at 90°

26 In the Duochrome test

A the Snellen letters used are red or green, while the background is black
B the subject should look at the letters for at least a minute before deciding which line is clearer
C it cannot be used for a red–green colour blind individual
D a myopic eye sees the 'red' letters more clearly than the 'green' letters
E it is especially useful before retinoscopy when the approximate refraction of the eye is unknown

27 Examples of optical aberrations is/are

A spherical aberration
B cylindrical aberration
C oblique astigmatism
D coma
E amblyopia

28 In subjective methods of refraction

A it is acceptable to find the best vision sphere before the power and axis of the trial cylinder is verified
B the cylindrical axis should be checked each time the cylindrical power is altered
C if rotation of the stenopaeic slit with the optimal spherical correction results in improvement in visual acuity in one meridian, astigmatism is present
D the stenopaeic slit measures 20 mm × 5 mm
E the methods can be omitted altogether if one is confident of the retinoscopy findings

29 A pin hole aperture

A totally corrects any refractive error
B causes increased refraction
C causes increased diffraction
D increases the depth of focus
E reduces spherical aberration

30 In refraction

A objective tests are used before subjective tests
B objective tests include retinoscopy
C subjective tests include keratometry
D subjective tests include astigmatic fans
E objective tests include the cross-cylinder

31 Retinoscopy

A if performed properly, can yield a degree of accuracy to the order of 0.25D in power
B if performed properly, can yield a degree of accuracy to the order of 5° in the cylindrical axis in astigmatism
C should be performed in a darkened room
D should ideally be performed by the examiner's right eye when the subject's right eye is examined
E reveals astigmatism if the points of neutralisation differ in different meridians

32 Fogging techniques in subjective refraction

A aim to encourage accommodation of the eye
B are especially important in the pseudophakics
C are especially important in the elderly
D may involve the use of a positive spherical lens
E should accompany the use of cycloplegic eye drops

33 If scissors shadow is encountered in retinoscopy

A it can be due to corneal scarring
B it can be due to posterior staphyloma
C an approximate correction may be found by finding the lens which causes the two bands of light to meet in the centre of the pupil
D it may be partially eliminated by altering the subject's gaze
E it implies the presence of regular astigmatism

34 The following is/are true regarding the direct ophthalmoscope.

A It contains a prism.
B It is useful for examining the anterior part of the eye.
C A hole is present in the mirror.
D The axis of observation should be in line with the corneal reflection of bulb image.
E It sometimes contains a green filter.

35 Use of the direct ophthalmoscope compared with the indirect ophthalmoscope

A gives a wider field of view
B is less affected by subject's state of refraction
C achieves a larger magnification
D is easier for examining retinal detachment
E is more useful in detecting eccentric fixation

36 The following statements about prismatic power is/are correct.

A An object placed at a distance of 1 m appears displaced by 1 mm by a prism of 1 Δ.
B An object placed at a distance of 2 m appears displaced by 1 cm by a prism of 2 Δ.
C An object placed at a distance of 1 m appears displaced by 2 cm by a prism of 2 Δ.
D An object placed at a distance of 10 cm appears displaced by 2 mm by a prism of 2 Δ.
E An object placed at a distance of 5 m appears displaced by 0.25 m by a prism of 5 Δ.

37 It is required to prescribe prisms of power equivalent to 3 prism dioptres base down and 4 prism dioptres base out, and a spherical correction of +10D for the right eye.

A The best option is to use two separate prisms.
B If one prism is used, the prism should be placed at 45° to the horizontal.
C If one prism is used, the required prismatic power is 5 Δ.
D If decentration is used, the lens should be decentred for 1.5 cm.
E If one prism is used, the required prismatic power is dependent on the refractive index of the material of the prism.

38 The focimeter

A can be used to measure the astigmatic axis of a lens
B can be used to measure the astigmatic power of a lens
C cannot measure the power of a prism
D consists of a telescope system
E contains targets (in the form of small dots) placed in the eyepiece

39 A Maddox rod

A is useful in measuring the near point of an eye
B can be used in detecting horizontal extraocular muscle imbalance
C consists of a series of low-powered plus cylinders
D can be used in detecting cyclophoria
E forms real images close to the eye

40 The following is/are true regarding measurement of intra-ocular pressure with an applanation tonometer.

A The force required to produce a constant area of contact between the tonometer head and the cornea is measured.
B When used correctly, the area of contact between the tonometer head and the cornea is 5 mm in diameter.
C The effect of surface tension increases as the area of contact increases.
D Readings of intra-ocular pressure should be taken when the outer edges of the two half circles are aligned.
E The tonometer head contains two prisms.

41 The following is/are true regarding the Hirschberg test.

A It is a subjective test of measuring the size of a squint.
B It is an accurate method of measuring the size of a squint.
C It is useful in both convergent and divergent squints.
D The angle of squint is about 30° if the reflex is at the temporal border of the pupil.
E The angle of squint is about 45° if the reflex is at the limbus.

42 Regarding a bifocal lens

A it can be used for presbyopia
B a Franklin bifocal lens (the near and distance portions were cut out from separate lenses) is most commonly used
C solid bifocal lenses are made of materials of different refractive indices
D the prismatic jump is minimum when the optical centre of each portion coincides with the distance and near visual point
E it is less likely to be tolerated by subjects with anisometropia

43 Iseikonic lenses

A are expensive to make
B can have magnifying power without refractive power
C are an example of an afocal lens with magnification
D can have a magnification up to 15%
E can be made to have magnification in one meridian only

44 A telescope system contains two convex lenses of 25D and 5D.

A The eyepiece is the 25D lens.
B They are separated by a distance of 20 cm.
C The image is erect.
D It is a Galilean system.
E The angular magnification is 5 times.

45 The following are advantages of close circuit television technique over spectacle magnifiers.

A Cheap.
B Adjustable magnification.
C Portable.
D Enhanced contrast.
E Normal reading position.

46 In the fitting of contact lenses

A the flatter the subject's cornea, the smaller the lens diameter necessary
B the smaller the lens, the greater the central separation (vaulting) between the lens and the cornea
C *either* the refraction *or* a measure of the corneal curvature is required before fitting
D peripheral curves are flatter than the central posterior curve
E a trial lens could be used

47 Soft contact lenses

A are usually made from poly-methylmethacrylate (PMMA)
B readily hydrate
C are pliable when they are dehydrated
D allow adequate oxygen to diffuse through the lens if the lens is thin
E can be used to deliver medication

48 Bifocal contact lenses

A may comprise an upper distance lens and a lower near lens
B may comprise a central distance lens and a peripheral near lens
C may comprise of distance correction in the left eye and near correction in the right eye
D are always preferred by users compared with bifocal spectacles
E always require stabilisation to prevent rotation

49 The following equipment is/are useful in checking contact lenses.

A A microscope with magnification of ×20.
B A radiuscope.
C Ultrasound sagittometer.
D Thickness gauges.
E Hruby lens.

50 The following is/are desirable physical properties of contact lens material.

A Low gas permeability.
B Biologically inert.
C Little frictional effect between the surface of the material and eye tissues.
D It takes part in enzymatic activity.
E Strong molecular adhesive force.

51 Ring scotoma

A is caused by prismatic deviation of high-power spectacle lenses
B consists of an area of central scotoma
C is particularly associated with correction of aphakia
D is stationary when the eye moves around
E results in 'jack in the box' phenomenon

52 Relative spectacle magnification

A by contact lens of axial myopia is less than 1
B in refractive myopia by a concave lens 15 mm from the eye is less than 1
C by contact lens of axial hypermetropia is less than 1
D of aphakics by contact lenses is about 1.1
E is the ratio of corrected image size to uncorrected image size

53 Relative accommodation

A is the difference between accommodation in the two eyes
B is the ratio of accommodation of an individual to that of a neonate
C is the amount of accommodation possible while convergence remains constant
D always has equal positive and negative portions
E remains constant while convergence varies

54 Regarding decentration of spectacle lenses

A decentration of 5 mm of a 10D lens would cause a prismatic effect of 5 Δ
B decentration of 1 mm of a 10D lens would cause a prismatic effect of 10 Δ
C decentration of 10 mm of a 5D lens would cause a prismatic effect of 5 Δ
D it sometimes serves useful purposes
E it is more pronounced in high powered lenses

55 In the detection of squints

A the cover test may reveal squints not apparent in the alternative cover test
B the alternative cover test may reveal squints not apparent in the cover test
C the Maddox rod may be used to detect cyclophoria
D the degree of a squint may be measured by a prism and a Maddox rod
E cyclophoria may be better detected with two Maddox rods (one before each eye) than one Maddox rod only

56 Asthenopia

A is caused by continuous accommodative effort
B is more likely to occur in myopia than hypermetropia
C is more likely to occur in 2D hypermetropia than 0DS/+2D DC × 45°
D does not occur if the amount of astigmatism is small
E occurs in anisometropic subjects more than pure hypermetropic subjects in general

57 Regarding materials used in spectacle lenses

A flint glass has a higher refractive index than crown glass
B flint glass is sometimes used for achromatic lenses
C plastic lenses are lighter than glass lenses
D plastic (Perspex) lenses are more susceptible to scratching than glass lenses
E plastic lenses warm up more rapidly than glass lenses

58 Accommodative effort needed to read a near object

A is unaffected whether a myope is corrected by contact lenses or by spectacles
B is less in a hypermetropic subject if corrected by contact lenses than by spectacles
C is an important consideration when prescribing for a presbyopic subject
D is lower for a myopic subject with than without spectacles
E is lower for a hypermetropic subject with than without spectacles

59 In the design of best form lenses

A the most important aberration to eliminate is chromatic aberration
B the more concave surface should be placed closer to the eye
C the base curve is the steeper curve
D for convex lenses of higher refractive power than 8D, no spherical lens can eliminate oblique astigmatism
E the base curve of a positive meniscus is the back surface

60 Keratometry

A measures the astigmatism of the eye
B is useful in fitting contact lenses
C measures the astigmatism of the curvature of the anterior surface of the cornea
D is useful before an intra-ocular implant
E is an alternative to retinoscopy

TEST 1
Answers

1 **A** = True **B** = False **C** = True **D** = True **E** = True

Visible light is only a very small part of the electromagnetic spectrum. All 'radiation' from the electromagnetic spectrum travels at the same velocity, about 300 million m s^{-1}. However, the 'radiations' have different wavelength and frequency. Since velocity = wavelength × frequency, the higher the frequency, the shorter the wavelength.

The 'radiations' with the higher frequency (shorter wavelength) than visible light include ultraviolet light, gamma rays and X-rays. The 'radiations' with lower frequency (longer wavelength) than visible light include infrared, microwaves, TV and radio waves.

Visible light is light detectable by the normal human retina. The wavelength is between 400 nm (blue light) and about 760 nm (red light). Sunlight contains a broad range of electromagnetic radiation, including ultraviolet, visible light and infrared. This can be easily demonstrated by dispersion of sunlight with a triangular prism.

The crystalline lens of the eye helps to reduce the amount of ultraviolet (wavelength below 400 nm) from entering the eye. The pigment epithelium of the iris is effective in absorbing radiation of light of all wavelengths, and helps to prevent reflection of light inside the eye.

2 **A** = False **B** = True **C** = True **D** = True **E** = True

The refractive index of a medium is a measure of its optical density, the higher the value, the more optically dense it is. For example, the refractive index of air is 1, that of water is 1.33, and that of crown glass is 1.5.

The **absolute** refractive index of a medium, *n*, is defined as

$$\frac{\text{Velocity of light in vacuum}}{\text{Velocity of light in medium}}$$

The velocity of light in vacuum is almost the same as that in air. Hence, the refractive index of air is 1.

When light travels from a less dense to a denser medium (e.g. from air to glass), the ray is deviated towards the normal. Conversely, when light travels from a denser to a less dense medium, it is deviated away from the normal.

Snell's Law states that when light travels from medium A to medium B,

a. the ray of incidence, ray of refraction and the normal are on the same plane,

b.

$$\frac{\sin i}{\sin r} = \frac{\text{refractive index of medium B}}{\text{refractive index of medium A}} \; \frac{\text{Velocity of light in medium A}}{\text{Velocity of light in medium B}}$$

where *i* is the angle of incidence, and *r* is the angle of refraction.

When light travels from a denser to a less dense medium, *r* is greater than *i*. The critical angle is the angle of incidence when the refracted angle is 90° (i.e. parallel to the junction between the two media). If the angle of incidence exceeds the critical angle, total internal reflection occurs (see Fig. 1.1).

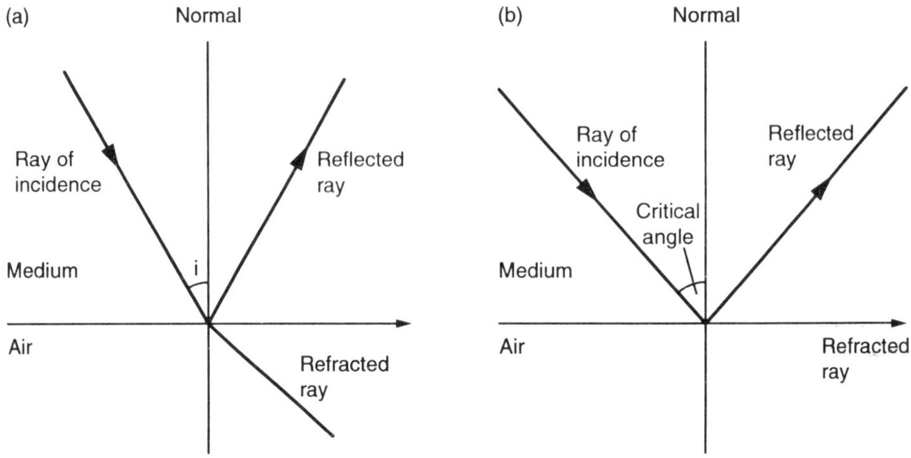

Figure 1.1 Refraction of light travelling from one medium to a denser medium with different angles of incidence.

3 **A** = False **B** = False **C** = True **D** = False **E** = True

Diffraction is the interference of the main wavefront formed by an aperture by secondary waves formed at the edge of the aperture. The smaller the aperture, the greater will be the effects at the edge and hence more diffraction will result. Diffraction is the property of waves, not particles. Diffraction can become important when a small lens is used. Diffraction limits the visual acuity obtained by a pinhole. Hence, the pinhole visual acuity is slightly below the visual acuity corrected by lenses.

4 **A** = True **B** = True **C** = True **D** = False **E** = False

When a ray of light enters a parallel glass block, the emergent ray is slightly less intense than the incident ray as there is some reflection at the first surface. The emergent ray is always parallel to the incident ray, and the displacement of the emergent ray increases as the width of the glass block increases (see Fig. 1.2). The displacement is more if the incident angle increases. Total internal reflection never occurs on the second surface, as the incident ray and emergent rays are always parallel, and hence the angle of incidence at the second surface can never exceed the critical angle.

5 **A** = False **B** = False **C** = True **D** = True **E** = True

See also Test 3 Question 5. The reduced eye is a simplified version of the

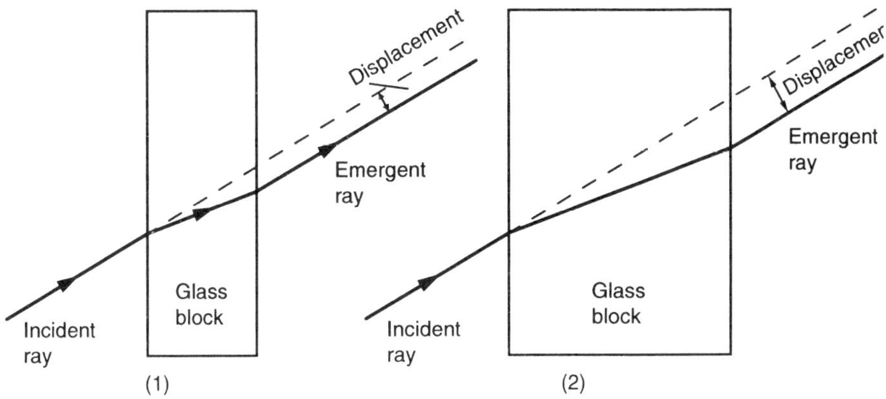

(1) The emergent ray is always parallel to the incident ray
(2) Displacement of the emergent ray increases with the thickness of the glass block

Figure 1.2 The relationship between displacement of an emergent ray and the thickness of a glass block.

schematic eye. A single principal point between the two principal points, and a single nodal point midway between the two nodal points, are selected. The eye is assumed to have a single refractive surface of power about 59D. The first focal point is 15 mm in front of the cornea, and the second focal point is at the retina (about 24.1 mm from the cornea). The nodal point lies roughly on the posterior surface of the crystalline lens, and is between the principal point and the back of the retina.

6 **A** = False **B** = True **C** = False **D** = True **E** = False

The general formula for transposing lens is:

X DS/YDC at *r* degrees is equivalent to

(X + Y) DS/–YDC at (*r* + 90)° if *r* is less than 90°
(X + Y) DS/–YDC at (*r* – 90)° if *r* is between 90 to 180°

In **A**, the correct transposition of +3DS/+2DC at 60° is +5DS/–2DC at 150°.
In **C**, the correct transposition of –4DS/+2DC at 30° is –2DS/–2DC at 120°.
In **E**, the correct transposition of +2DS/–1DC at 90° is +1DS/+1DC at 90°.

7 **A** = True **B** = True **C** = False **D** = False **E** = False

A cross-cylinder is a sphero-cylindrical lens with one meridian of equal power but the opposite sign to the other meridian. It may increase the interval of Sturm if the cylindrical axis of the same sign as the conoid of Sturm is in alignment. It may decrease the interval of Sturm if the cylindrical axis of the opposite sign as the conoid of Sturm is in alignment. The position of the circle of least diffusion is not affected by addition of a cross-cylinder, as it would add the same refractive power of one focal line and reduce the refractive power of the other focal line. Since the circle of least diffusion lies dioptrically midway between the two focal lines, its position is unchanged.

8 **A** = False **B** = True **C** = False **D** = True **E** = True

The accommodative–convergence:accommodation ratio is, on the whole, constant for the same individual, but varies amongst different individuals. It is a measurement of the effect of accommodation on convergence (AC:A ratio), and can be determined clinically. This is done by measuring the change in lateral phoria on changing the accommodation demand (by adding either convex or concave lenses of known power). The average ratio in the population is about 3.5 to 4 Δ of convergence to each D of accommodation. In mild prebyopia, it is useful to determine the AC:A ratio. If the ratio is high, the add should be minimised. If the ratio is low, the add can be made stronger.

9 **A** = False **B** = False **C** = False **D** = True **E** = False

When one observes a cross through a concave lens, a 'with movement' is seen. An 'against movement' is observed through a convex lens. In both cases, no distortion is observed.

When an astigmatic lens is used, there is distortion. Rotation of the lens causes a 'scissoring' effect with the images of the two arms of the cross rotating in different directions.

When a prismatic lens is used, there is constant displacement of the cross with no distortion of the cross.

10 **A** = False **B** = False **C** = False **D** = True **E** = False

In regular astigmatism, the principal meridians are at right angles to each other, and it can be totally corrected by cylinders. In irregular astigmatism, the principal meridians are not at right angles to each other, and it cannot be totally corrected by cylinders. Symmetrical astigmatism occurs when the principal meridians of each eye are symmetrical with the median vertical line as the axis of symmetry. In practice, symmetrical astigmatism occurs when the principal meridians of each eye corrected by cylinders of the same sign add up to about 180°. Symmetrical astigmatism is more common than asymmetrical astigmatism, and is less likely to cause head tilt or torticollis.

11 **A** = True **B** = False **C** = False **D** = True **E** = False

If a correcting lens of focal length f metres focuses light on the retina, then a correcting lens of focal length $(f + d)$ metres is required if the lens is placed d metres further away. Since refractive power = 1/focal length, it can be proved mathematically that,

$$F_d = \frac{F}{1 + dF}$$

where F_d = refractive power of the correcting lens placed d metres away from the eye (negative if moved towards the eye), F = refractive power of the correcting lens of the original lens in dioptres (+ve for convex lens, –ve for concave lens) and d = distance moved away from eye (negative if moved towards eye).

In general, a weaker convex lens or a stronger concave lens is required if the correcting lens is moved away from the eye.

In the question, for a spectacle lens with a back vertex distance (BVD) of 15 mm (0.015 m)

$$F = 14\text{D}$$
$$d = 0.015 \text{ m}$$

$$F_d = 14/(1 + 0.015 \times 14)$$
$$= 11.57\text{D}$$

Likewise, for a spectacle lens with BVD of 12 mm (0.012 m),

$$F_d = 14/(1 + 0.012 \times 14)$$
$$= 12\text{D}$$

For a spectacle lens with BVD of 10 mm (0.01 m)

$$F_d = 14/(1 + 0.01 \times 14)$$
$$= 12.28\text{D}.$$

12 **A** = True **B** = False **C** = True **D** = True **E** = True

The far point of the eye is the position of an object where image would fall on the retina of an eye in the absence of accommodation. A total of $1/y$ dioptre is required to bring an image y metres away from the eye. In the hypermetropic eye, there is excess refractive power in the eye, and even objects at infinity would form an image behind the retina. Accommodation is required to focus the image on the retina. Hence, the far point is virtual. In a myopic eye of x dioptre, an object $1/x$ metre away from the eye would form an image on the retina. In the emmetropic eye, the far point is at infinity.

13 **A** = False **B** = False **C** = False **D** = False **E** = True

In myopia, light from distant objects (parallel light) enters the eye and converges at a point in front of the retina (see Fig. 1.3).

Myopia may be caused by an abnormally short eyeball (axial myopia), or an abnormally strong refractive medium in the eye (refractive myopia). Axial myopia is the most common cause, and may be hereditary. Refractive myopia includes conditions caused by abnormally high corneal refractive power (e.g. keratoconus) or abnormally high refractive power in the lens (e.g. nucleus sclerosis).

Myopia should be corrected with a concave lens.

The further the concave lens from the eye, the higher the power of the lens needed.

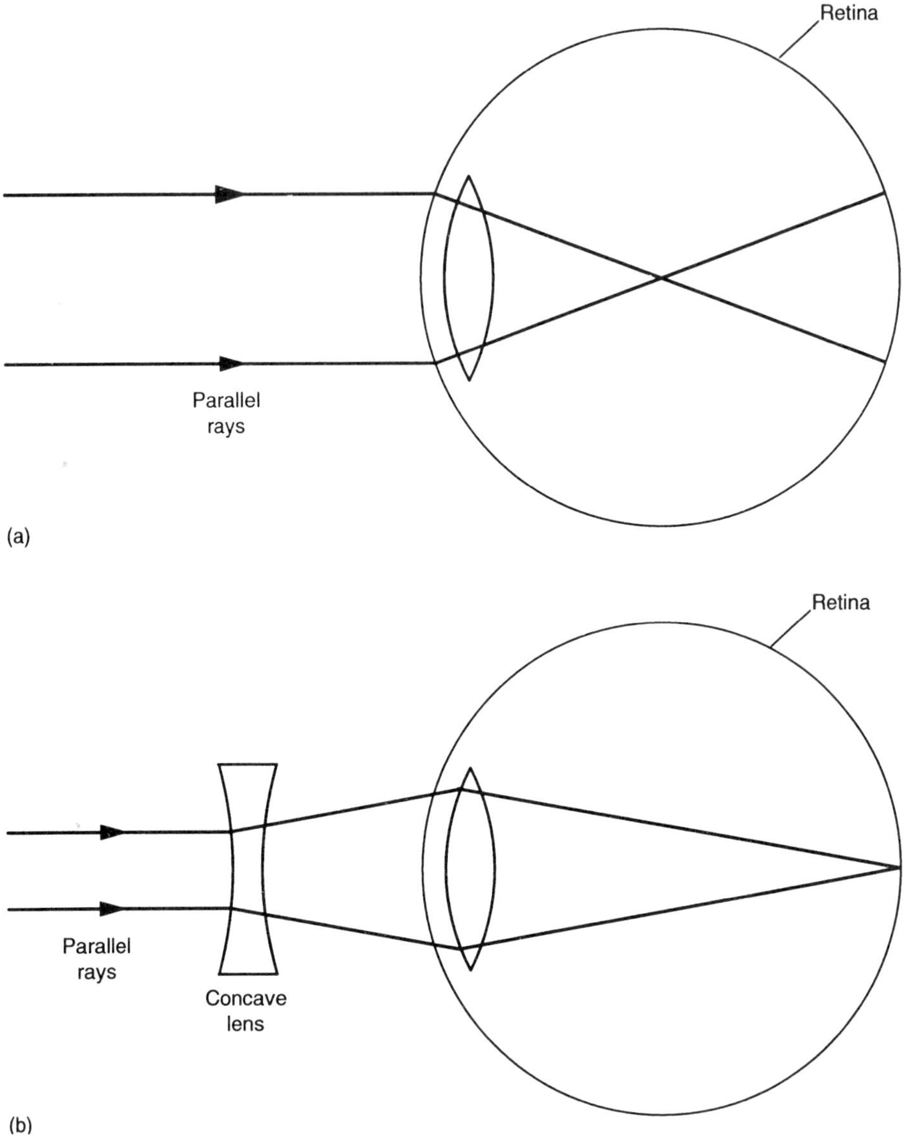

Figure 1.3 Correction of myopia by concave lens. (a) Myopia; (b) myopia corrected by a concave lens.

14 **A** = False **B** = True **C** = True **D** = False **E** = True

Latent hypermetropia is the portion of hypermetropia compensated by ciliary muscle tone, which can only be revealed with cycloplegics such as atropine. The rest of the hypermetropia is called **manifest hypermetropia**. Of the manifest hypermetropia, the portion which can be reduced by accommodation is called the **facultative hypermetropia**, and the portion which cannot is called **absolute hypermetropia**.

Hypermetropia

Total = latent + manifest
Manifest = facultative + absolute

In infancy, the ciliary muscle tone is high, so that most of the hypermetropia is latent. This can only be revealed by cycloplegics. As age increases, the latent hypermetropia decreases while the manifest hypermetropia increases. As the amplitude of accommodation decreases with age, most hypermetropia becomes absolute at about 60 years.

15 **A** = False **B** = True **C** = False **D** = True **E** = True

Astigmatism needs correction if the visual acuity is significantly reduced or if there are symptoms of asthenopia. An asymptomatic simple astigmatism of 0.25D does not require optical correction. Astigmatism should be corrected for both near and distant vision, as accommodation cannot correct for the astigmatism. Compound hypermetropic astigmatism (where the foci for both meridian fall behind the retina) causes more reduction in visual acuity than mixed astigmatism (where the focus for one meridian falls behind the retina and the other falls in front). This is because, in mixed astigmatism, the circle of least diffusion (which lies between the two foci) is likely to be nearer to the retina.

16 **A** = False **B** = True **C** = True **D** = True **E** = True

When the normal eye accommodates, the pupil becomes smaller, and the eyes converge. The convergence can cause convergent squints in those with abnormally high AC/A (accommodative convergence to accommodation) ratio.

Accommodation occurs equally in both eyes at the same time, so that it cannot compensate for unequal hypermetropia.

17 **A** = True **B** = True **C** = False **D** = True **E** = True

Pathological axial myopia is thought to be genetic in origin, and is more common in some ethnic groups such as Japanese and Chinese. The refractive error usually appears in childhood and can increase progressively beyond adolescence to 20D or more. Marked degenerative changes are characteristic features, although they rarely appear in adolescence.

18 **A** = False **B** = False **C** = False **D** = True **E** = True

The accommodation of the eye is its ability to increase the refractive power of its lens. The crystalline lens is held under tension by the suspensory ligament, which attaches it to the ciliary muscle. There are two opposite forces determining the shape: the elasticity of the capsule which maintains the lens in its accommodated shape, and the elasticity of the lens which resists this. Accommodation is caused by the following events:

Ciliary muscle contraction	→	Ciliary body approaches lens
	→	Decrease in tension on suspensory ligament (zonules)
	→	Less tension on capsule
	→	Lens (especially anterior surface) assumes a globular shape
	→	Increase in refractive power of the lens

The amplitude of accommodation is about 14D at birth, and declines constantly with age. The range of accommodation is the distance between the far point and the near point. The near point gradually increases with age, although very slowly in the first decade. In an emmetropic eye with accommodation of 12D, the near point is 8.3 cm. When the accommodation is reduced to 4D, the near point is 25 cm. However, it does not become symptomatic until the near point recedes to a point beyond comfortable reading distance. All cycloplegic drugs (e.g. Tropicamide, cyclopentolate, atropine) paralyse the ciliary muscle, and hence reduce the amplitude of accommodation.

19 **A** = True **B** = False **C** = False **D** = True **E** = True

The near point is the nearest point at which objects can be clearly seen without blurring. At this point, maximum possible accommodation has been used, and objects closer to the eye cannot be seen clearly. In myopia compared with emmetropia, there is excessive refractive power of the eye, the image is in front of the retina in the absence of accommodation, and objects closer to the eye can be focused on the retina. Conversely, in hypermetropia, there is a relative deficit in refractive power of the eye. The image is behind the retina in the absence of accommodation, and some accommodation is used to overcome this. Hence, the near point is further away. As age increases, the nucleus of the lens becomes harder and less plastic, and the amplitude of accommodation decreases. Hence, the near point moves further away.

20 **A** = False **B** = True **C** = False **D** = False **E** = False

A Fresnel prism consists of many prisms contained in a strip of plastic about 2 mm thick. It may be used to relieve symptoms of diplopia in selected patients. It can be pasted onto spectacle lenses, and is therefore especially suitable if it is uncertain whether prisms are required or not. The multiple prisms have axes parallel to one another, and produce the refracting power equivalent to a single large prism with an identical refracting angle. Porro and Dove prisms are used for inverting images.

21 **A** = True **B** = True **C** = True **D** = False **E** = False

Total internal reflection occurs when light enters from one medium to a less dense medium, at an angle exceeding the critical angle. There is no refraction, and all the light is reflected.

Glass and plastics can be made into fibres, smaller than 50 μm thick. If light is made to enter one end of a fibre at an appropriate angle, it will undergo a series of total internal reflection, and travel down the fibre. A bundle of closely packed fibres with their relative positions maintained, can be used in endoscopy. Optical fibres are also used in telephone communications, and their advantages are that information travels at the speed of light, and there is little loss through the sides.

The primary rainbow is caused by refraction of the light as it enters a drop, with total internal reflection at the back surface, and finally refraction as the light emerges from the drop. A secondary rainbow is associated with two internal reflections, and the order of the colour is reversed.

The triangular prism can be used as a mirror when light enters it exceeds the critical angle. It is used extensively in optical instruments including binoculars.

An antireflection coating of a lens uses the principle of interference of light. A transparent material is evaporated onto a lens to a thickness of a quarter of the wavelength. Some light of this wavelength will be reflected from this surface. If the reflection from the glass is equal in intensity but 180° out of phase to this, the reflections will cancel each other resulting in no reflection and 100% refraction

22 **A** = True **B** = True **C** = False **D** = True **E** = True

Correction of aphakia by spectacles may cause numerous problems, due to the strength of the lens and the absence of accommodation of the eye. Examples are as follows:

a. prismatic effects, causing ring scotoma and the 'jack in the box' effect;
b. spherical aberration;
c. image distortion with pin-cushion effects;
d. reduced visual fields;
e. magnification of the retinal image of about 30%;
f. loss of binocular single vision;
g. sensitivity of the power of spectacle correction required to small changes of the back vertex distance.

23 **A** = False **B** = False **C** = True **D** = False **E** = False

Back vertex distance is the distance between the eye and the back surface of the spectacle lens. It is different for each individual and for different spectacle frames, owing to different facial contours. In practice, the difference in back vertex distance makes a significant difference to the power of the lens prescribed only if the power of the lens exceeds 5D. It is especially important in aphakia, as the power of the lens prescribed is likely to be around 11D.

24 **A** = True **B** = False **C** = True **D** = False **E** = True

Visual acuity is usually measured by resolution—the ability to discriminate two spatially separate targets. It measures the minimum angle two targets subtend at the nodal point of the eye before they can be seen as separate. It is a measure of the function of the cones in the fovea. Clinical tests include the Snellen chart or the Landolt ring. Snellen's chart depends on the literacy of the subject. However, it is less confusing for both the subject and the examiner if the subject has achieved minimal literacy.

25 **A** = True **B** = False **C** = True **D** = False **E** = False

The cross-cylinder is a sphero-cylindrical lens with the spherical power half that of its cylindrical power. Examples are 0.5DS/–1.0DC @90° or 0.25DS/–0.5DC @90°. It is the same as superimposing two cylindrical lenses of equal power but opposite sign with their axis at 90°.

To check the **axis** of the trial cylinder, place the **handle** of the cross-cylinder in alignment with the axis of the trial cylinder. Compare the visual effect when the cross-cylinder is flipped over by turning the handle. To compare the **power** of the trial cylinder, place the axis of the cross-cylinder in alignment with the axis of the trial cylinder, and compare the visual effect when the cross-cylinder is flipped over by turning the handle.

26 **A** = False **B** = False **C** = False **D** = True **E** = False

In the Duochrome test, the Snellen letters are black while the illumination background is red or green. The subject has to decide which row of letters appears clearer.

The subject must not look at the letters for too long. Focusing on the letters for too long would encourage accommodation of the lens in the eye, which would give a refractive result indicating the eye being more myopic than it actually is.

This test does not depend on the recognition of colour, and it can therefore be used for the red–green colour blind.

The test makes use of the chromatic aberration of the eye. The green (shorter wavelength) light is focused in front and the red (longer wavelength) is focused behind the retina in an emmetropic eye. Hence, the slightly myopic eye would see the red letters more clearly than the green letters. Conversely, the slightly hypermetropic eye would see the green letters more clearly than the red letters.

The test can only be used if the eye is nearly emmetropic. Hence, it cannot be used before the approximate refractive error of the eye is known.

27 **A** = True **B** = False **C** = True **D** = True **E** = False

Optical aberrations are ways which cause refracting systems, including spectacle lenses and the eye, to deviate from perfect theoretical models. Spherical aberration is caused by the periphery of a lens having a more prismatic effect than the centre. Coma is due to the same mechanism but from points not on the optical axis of the lens, sometimes causing a 'coma shaped' image.

Oblique astigmatism occurs when light enters a spherical lens obliquely, so that the focal length for the vertical and horizontal meridians differ from each other. In amblyopia, the optical system of the eye is normal.

28 **A** = True **B** = True **C** = True **D** = False **E** = False

The spherical correction is usually estimated first by finding the best vision sphere (together with the cylinder found on retinoscopy) which would give the best visual acuity. Then, the cylindrical axis and power are verified. It is important to verify the axis again each time after the power is changed. A sphere of half the power of the added cylinder of the opposite sign should be added each time a cylinder is added. The sphere should again be verified after the cylindrical axis and power are verified.

Subjective methods of refraction should always be performed even if one is confident of the retinoscopy findings, as the subject may have an unusual accommodation pattern in normal vision.

The stenopaeic slit measures 20 mm × 1 mm. If rotation of the stenopaeic slit with the optimal spherical correction results in visual acuity improvement in one meridian, then astigmatism is present in that axis.

29 **A** = False **B** = False **C** = True **D** = True **E** = True

The pinhole does not correct all refractive errors, as diffraction would limit the visual acuity. It increases diffraction, as the secondary wavelengths from the edges of the pinhole become more important. The depth of focus is the degree that an image can be out of focus without much blurring. A pinhole aperture increases the depth of focus. It also reduces spherical aberration as the aberration is maximum at the periphery.

30 **A** = True **B** = True **C** = False **D** = True **E** = False

Objective tests in refraction are those that the examiner can determine without responses from the subject. This includes retinoscopy, keratometry and automatic refractometry. Subjective tests in refraction are those that require responses from the subject, and include use of the cross-cylinder, the Duochrome tests, astigmatic fans and blocks. Objective tests should be performed first to determine the approximate refractive error, and subjective tests should then be carried out to verify any adjustments made if necessary.

31 **A** = True **B** = True **C** = True **D** = True **E** = True

Retinoscopy is an accurate method of objective refraction if the media of the subject's eye are clear and the examiner is skilled. It can be accurate in ideal circumstances to 0.25D in power and 5° in the cylindrical axis. It is easier if performed in a darkened room and to examine the subject's right eye with the examiner's right eye. If the points of reversal are the same in all meridians, then there is only spherical correction. If they are different for two different meridians, then astigmatism is present.

32 **A** = False **B** = False **C** = False **D** = True **E** = False

Fogging techniques in subjective refraction aim to discourage accommodation of the eye in order to give an accurate measurement of the refractive error of the eye. This is achieved by placing a positive spherical lens (a convex lens) in front of the eye until the distant visual acuity is 6/60. As there is no accommodation in the pseudophakic and the elderly, these techniques need not be used. Cycloplegics should not be used in subjective refraction.

33 **A** = True **B** = True **C** = True **D** = True **E** = False

Scissors shadow in retinoscopy occurs when the refractivity of one area of the pupil differs from the other. Near the point of reversal, one area of the pupil shows the phenomenon of myopia while the other shows the phenomenon of hypermetropia. It is usually caused by corneal scarring, and the blades of the scissors usually point to the cicatrix. It may also occur in lenticular irregular astigmatism or posterior staphyloma. In the latter case, the blades of the scissors would point to the posterior staphyloma. Scissoring phenomenon can sometimes be avoided by changing the working distance or asking the subject to look in a different direction. An approximate refractive error can be determined from the lens required for the two blades to meet in the centre of the pupil.

34 **A** = False **B** = True **C** = True **D** = False **E** = True

The direct ophthalmoscope illuminates a subject's retina by a mirror reflecting light from a bulb passing through a series of lenses. The mirror contains a hole through which the examiner can examine the subject's retina. The axis of observation should be slightly out of line with the corneal reflection of the bulb image, as the corneal image would otherwise interfere with the examination. Most direct ophthalmoscopes contain a green filter which can be selected to examine a structure containing blood, such as haemorrhages and blood vessels. The direct ophthalmoscope can also be used to examine structures at the anterior part of the eye by selecting convex lenses from the ophthalmoscope.

35 **A** = False **B** = False **C** = True **D** = False **E** = True

The advantages of the direct ophthalmoscope are that it is portable, cheap, can be run by battery, has a higher magnification (about 15 times for an emmetropic eye), and the image is erect and easily interpreted. The disadvantages are that it is difficult to visualise the peripheral fundus, the field of view is small, the view varies depending on the subject's refractive state, and the view is not stereoscopic. An indirect ophthalmoscope is more expensive and bulky, and proves intolerable to some patients due to its brightness. On the other hand, good illumination allows the fundus to be seen when there are opacities in the media. The magnification is smaller (about three times for a 20D lens, and five times for a 13D lens), but the field of view is larger. Also, stereoscopic view is achieved. It is best used to view the peripheral fundus as well as the retinal hole and detachment. It is indispensable during a retinal detachment operation.

36 **A** = False **B** = False **C** = True **D** = True **E** = True

A prism is said to have a power of 1 Δ if objects 1 m away appear to be displaced by 1 cm. In general, if an object *x* metres away appears to be displaced by *y* cm, the power of the prism can be represented by (*y/x*) prism dioptres.
In **A**, the prismatic power is 0.1/1 = 0.1 Δ.
In **B**, the prismatic power is 1/2 = 0.5 Δ.
In **C**, the prismatic power is 2/1 = 2 Δ.
In **D**, the prismatic power is 0.2/0.1 = 2 Δ.
In **E**, the prismatic power is 25/5 = 5 Δ.

37 **A** = False **B** = False **C** = True **D** = False **E** = False

An obliquely placed prism (base down and out) can be more conveniently used than two separate prisms. Alternatively, the lens can be decentred.

The power of the prism required can be calculated by the following diagram:

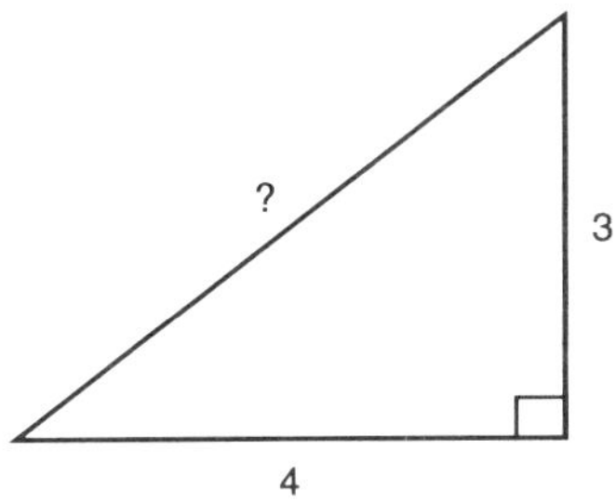

By Pythagoras' Theorem, the prismatic power required is $\sqrt{(3^2 + 4^2)} = 5$ Δ.

Using the formula **Prismatic power** = **refractive power of lens** × **decentration** (in cm), the lens should be decentred by 10/5 = 2 cm.

38 **A** = True **B** = True **C** = False **D** = True **E** = False

The focimeter can be used to measure the spherical component of a lens, the astigmatic axis and power of a lens, and the power of a prism.

The focimeter consists of a light source with the target of a ring of small dots, a standard lens, and a telescope system. The instrument is first set to zero, when light entering the telescope system is parallel. The lens under test is placed on the lens rest so that the surface being tested faces the lens rest (and the standard lens). The target is moved so that it is once again under focus. The distance moved is related to the refractive power of the surface of the test lens, and is converted into dioptres.

39 **A** = False **B** = True **C** = False **D** = True **E** = True

A Maddox rod is used to measure the extraocular muscle imbalance, including horizontal and vertical muscle imbalance and cyclophoria. It consists of a series of high-powered positive cylinders with their axis parallel to one another, and mounted on a trial frame. It is usually tinted red in colour. It is placed close in front of the subject's eye. Each cylinder forms a real image close in front of the subject's eye. This forms a line of real images parallel to the axis of the cylinders.

They are too close to the eye to be focused upon, and the subject sees instead a line of virtual images perpendicular to the axis of the cylinders.

40 **A** = True **B** = False **C** = False **D** = False **E** = True

The applanation tonometer is used to measure the intra-ocular pressure. It consists of an tonometer head mounted on a lever, the tension of which can be adjusted with a knob. The tonometer head contains two prisms, with the apex and base in the opposite direction.

The intra-ocular pressure is taken as the force applied by the tonometer head on the cornea to create an area of contact of 3 mm in diameter. When the area of contact is 3 mm in diameter, the effect of surface tension and rigidity of the cornea cancel out. When the area is more than 3 mm in diameter, it has been found that the effect of corneal rigidity exceeds surface tension.

The examiner looks at the area of corneal contact, which is split into two half circles by lateral displacement of the prisms. When the inner edges of the meniscus are in contact with each other, the area of corneal contact is about 3 mm in diameter. The intra-ocular pressure can then be recorded.

41 **A** = False **B** = False **C** = True **D** = False **E** = True

The Hirschberg test is a very crude objective method of estimating the size of a squint. Light from a pen torch is directed at the eyes, and the position of the corneal reflex is noted. If the reflex is central in one eye, but at the temporal border of the pupil of the other, the size of the convergent squint is about 15°. If the reflex is at the limbus, the size of the convergent squint is about 45°.

42 **A** = True **B** = False **C** = False **D** = False **E** = True

The **Franklin bifocal lens** consists of two lenses (upper lens for distant vision, and lower lens for reading) cut from separate lenses put together on a spectacle frame. It is seldom used nowadays. The **cemented wafer** contains a principal distance lens with an additional 'wafer' cemented onto the distance lens. **Solid bifocals** are made by one material only, but the lower portion for near vision is ground to change its refractive power. **Fused bifocals** are made with a crown glass lens, with a portion of the lower part hollowed out and replaced by a material with a higher refractive index.

If the optical centre of the upper lens coincides with the distance visual point (point on a spectacle where the visual axis of the eye passes through) and the lower lens coincides with the near visual point, there will inevitably be prismatic jump at the boundary between the two lenses. There is no prismatic jump if the optical centre of each lens lies at the boundary of the two lenses. However, there will then be problems of shifting the optical centres of each lens away from the visual axis. The prismatic jump is greater the higher the power of the lens. Hence, it is unlikely to be tolerated by subjects with high refractive errors or when there is significant difference in refractive errors between the two eyes (anisometropia).

43 **A** = True **B** = True **C** = True **D** = False **E** = True

Iseikonic lenses are lenses with magnification power but no refractive power. They produce angular magnification by utilising the effect of the lens thickness and the refractive power of the front surface, which is neutralised by the refractive power of the other surface. They are examples of afocal lenses. Due to limitations on the thickness of a lens, the maximum magnification produced is about 5%.

They are expensive and require special equipment to be manufactured. They can either be made with equal magnification in all meridians, or with magnification in one meridian only.

44 **A** = True **B** = False **C** = False **D** = False **E** = True

The eyepiece has a higher dioptric power than the objective. The distance between the two convex lenses should be equal to the sum of their focal lengths. The focal lengths of the lenses are (1/25) m = 4 cm and (1/5) m = 20 cm. Hence, they should be separated by (20 + 4) = 24 cm so that the foci of the lenses coincide.

The angular magnification of the telescope equals the ratio of the focal lengths of the lenses. Hence, angular magnification = 20/4 = 5 times (see Fig. 1.4)

The image of this astronomic telescope system is virtual and inverted. In a Galilean system, the eyepiece is a concave lens.

45 **A** = False **B** = True **C** = False **D** = True **E** = True

The advantages of closed circuit television (CCTV) techniques are: higher and adjustable magnification (up to 80 times), the possibility of image reversal (e.g. white print on black background), CCTV allows enhanced contrast, and allows normal reading position. The disadvantages are that it is expensive and not portable.

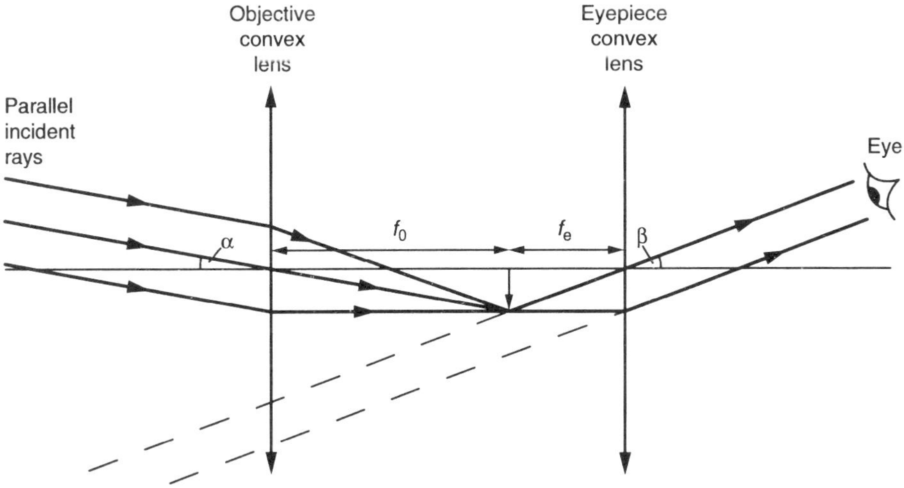

Figure 1.4 The astronomic telescope. f_0 = focal length of the objective lens, f_e = focal length of the eyepiece.

46 **A** = False **B** = False **C** = False **D** = True **E** = True

The basic measurements of a contact lens are the central posterior curve (base curve), intermediate and peripheral curves, refractive power and the thickness. The intermediate curve is flatter than the central posterior curve, and the peripheral curve in turn is flatter than the intermediate curves. This is both to fit the aspherical surface of the cornea and to allow a good flow of tears below the lens.

The smaller the lens, the less will be the sagittal depth of the lens. Since the cornea is steeper in the centre, there will be less separation between the lens and the cornea. Hence, the smaller the lens, the smaller will be the central separation (vaulting) between the lens and the cornea.

Refraction would determine the power of the contact lens needed, and a measure of the corneal curvature would determine the central posterior curve (*K* value) needed. Hence, both refraction and corneal curvature are required for fitting of contact lenses.

A trial lens is a much more reliable method of fitting, and should be used before contact lenses are ordered.

47 **A** = False **B** = True **C** = False **D** = True **E** = True

PMMA (poly-methylmethacrylate) is only slightly hydrophilic and is used to make hard contact lenses. Soft contact lenses are generally made from PHEMA (poly-2-hydroxyethylmethacrylate) or VP (vinylpyrrolidone), both of which are hydrophilic. Soft contact lenses generally mould to the shape of the eye. In the pure state, they are rigid and brittle. However, they hydrate easily. In the hydrated state, they are pliable, and when deformed, quickly return to their original shape. If the lens is thin, there is enough oxygen supply to the corneal epithelium. Even when a lens is thick, there is enough tear circulation between the lens and the cornea for oxygen exchange. There is sometimes a problem with the build up of a drug in the contact lenses when medications are administered. On the other hand, it is sometimes a useful way of delivering medication (e.g. delivering Pilocarpine).

48 **A** = True **B** = True **C** = True **D** = False **E** = False

Bifocal contact lenses can be designed in three ways.

a. The distance portion is in the upper part and near portion is in the lower part of the lens.
b. The distance portion is in the centre and the near portion is in the periphery of the lens.
c. The distance correction is in one eye and near correction is in the other.

While bifocal contact lenses are accepted by some users, others find them impossible to adjust to and prefer bifocal spectacles.

If design a is used, rotation must be prevented by stabilisation. In design b and c, this is unnecessary, as rotation does not affect the visual result. Stabilisation for design a can be achieved either by prism ballast (attaching a prism at the base of the lens), by truncation (cutting off the edge of a lens) or both.

49 **A** = True **B** = True **C** = True **D** = True **E** = False

Microscopes with a magnification of ×20 are useful in examining for lens surface and contact lens edge defects. Radiuscopes can be used to measure the central and peripheral radii, and lens thickness. Ultrasound sagittometers can also be used to measure contact lens curvature, especially soft lenses. Thickness gauges are used to measure the thickness of contact lenses. A Hruby lens is used to visualise the posterior fundus only.

50 **A** = False **B** = True **C** = True **D** = False **E** = False

Materials used for contact lenses should be biologically inert, and should not take part in enzymatic properties. The material should not change its property within the physiological pH. It should have a high gas permeability to allow sufficient oxygen to diffuse into the cornea. It should have little frictional effect between the material and eye tissues so that trauma to the cornea is minimised.

51 **A** = True **B** = False **C** = True **D** = False **E** = True

Ring scotoma may occur when high-powered spectacles are worn, especially in aphakia. Light entering through the lens undergoes prismatic deviation, while light entering outside the lens is not deviated. Hence, there is a ring of visual field just outside the lens where it cannot enter the eye and be seen by the subject. This is the ring scotoma. The location of the scotoma moves with the movement of the eye, and results in the 'jack in the box' phenomenon.

52 **A** = False **B** = True **C** = True **D** = True **E** = False

Spectacle magnification is the ratio of corrected image size to uncorrected image size. **Relative** spectacle magnification is the ratio of corrected ametropic image size to emmetropic image size.

In refractive ametropia, the magnification/diminution can be thought of as a system of lenses similar to the Galilean telescope. The magnification equals the ratio (excessive refractive power of the eye/power of lens used). In refractive myopia, the power of the spectacle lens used is more than the actual degree of myopia in the eye, as the lenses are further away from the eye. Hence, there is always diminution unless the lens is at the nodal point of the eye. The nearer the correcting lens is to the eye, the less the diminution,

Aphakia is an example of refractive hypermetropia of about 15D. If it is corrected by a contact lens, it is placed about 7 mm in front of the nodal point. Applying the formula

$$F_d = \frac{F}{1 + dF}$$

where F_d = refractive power of the correcting lens placed d metres away from the eye (negative if moved towards the eye), F = refractive power of the correcting lens of the original lens in dioptres (+ve for convex lens, –ve for concave lens) and d = distance moved away from eye (negative if moved towards eye)

$$Fd = 15/(1 + 0.007 \times 15) = 13.6D$$

Hence, Relative Spectacle Magnification = 15/13.6 = 1.1.

In pure axial ametropia, the refractive power of the eye is normal, but the axial length of the eye is either elongated (in myopia) or shortened (in hypermetropia). If a thin lens is used to correct the ametropia, and the centre of the lens is in the anterior focal point of the eye, light entering the lens would pass parallel to the optical axis of the eye. Hence the image size would be the same irrespective of the axial length of the eye. **Knapp's Rule** states that in pure axial ametropia, correction by a lens placed at the anterior focal plane of the eye would produce the same size image as in emmetropia. In axial hypermetropia, moving the lens towards the eye from the anterior focal point would result in diminution (RSM <1), while moving away from the eye results in magnification (RSM > 1).

53 **A** = False **B** = False **C** = True **D** = False **E** = False

Relative accommodation is the amount of accommodation possible while convergence remains constant. It is a measure of the degree of dissociation between accommodation and convergence. The positive portion of the relative accommodation is the amount in excess of convergence, and is measured by the most powerful negative lens which can be placed in front of an eye for an object to be seen without blurring. The negative portion of the relative accommodation is the amount below that of convergence, and is measured by the most powerful positive lens which can be placed in front of an eye for an object to be seen without blurring. The positive portion of relative accommodation increases as an object moves towards the eye, and the negative portion increases as an object moves away from the eye.

54 **A** = True **B** = False **C** = True **D** = True **E** = True

Decentration may occur either by shifting the whole lens to one side so that the optical axis and the geometrical axis (i.e. the centre of the lens in the spectacle frame) do not coincide, or by cutting out a smaller lens from the periphery of a larger lens. The prismatic effect is given by the formula

Prismatic effect (in prism dioptre) = power of lens (in dioptres) × distance decentred (in cm)

The formula can be easily proved as follows.

Prismatic effect = Prismatic deviation (cm)/distance from lens (m)
= d/f
= dP where P is the power of the lens (since $P = 1/f$)

In **A**, prismatic effect = 0.5 cm × 10D = 5 Δ.
In **B**, prismatic effect = 0.1 cm × 10D = 1 Δ.
In **C**, prismatic effect = 1 cm × 5D = 5 Δ.

Decentration sometimes serves useful purposes, such as correction of

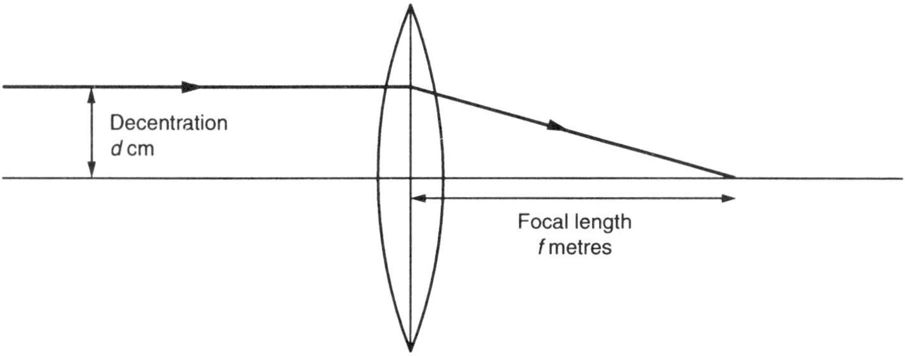

Figure 1.5 Decentration of lenses.

convergence excess or deficit, or when the subject's activities require vision not in the primary position (e.g. when playing snooker) (see Fig. 1.5)

55 **A** = False **B** = True **C** = True **D** = True **E** = True

The cover–uncover test will reveal any manifest squints present (tropia). However, detection of a latent squint requires the alternate cover test. In the alternate cover test, one eye is covered for about 2 s, and the cover is quickly shifted to the other eye. The uncovered eye is observed for movement.

In a horizontal or vertical squint, a prism can be used in conjunction with a Maddox rod to measure the squint. The power of the prism is varied so that the line seen through the Maddox rod goes through the white point seen with the other eye. The degree of squint can then be measured.

Cyclophoria may be detected by using two Maddox rods of different tint, one in each eye with the axis in the same direction. The images are compared. If the two line images are in different directions, one of the Maddox rods can be rotated until they coincide. The degree of cyclophoria can therefore be measured.

56 **A** = True **B** = False **C** = False **D** = False **E** = True

Asthenopia is caused by accommodative fatigue when there is continuous accommodative effort. It is more likely to occur in hypermetropia than myopia, as hypermetropic subjects need to exert more accommodation for near vision. In astigmatism, the image is always blurred, and there is always a stimulus to accommodative effort. Since accommodation cannot correct astigmatic errors, a small astigmatic error may continuously stimulate accommodative effort. Similarly, in anisometropic subjects, the accommodative effort of the two eyes cannot be dissociated. Hence, there is always a stimulus for accommodation, causing asthenopia.

57 **A** = True **B** = True **C** = True **D** = True **E** = True

Flint glass has a high refractive index of 1.62, and is used for making achromatic lenses or bifocal lenses. Plastic (Perspex) lenses are lighter and do not break or splinter easily. However, they can be scratched easily, and can warp under heat or pressure. This can be remedied by the use of a thermosetting resin. They also warm up more rapidly than glass lenses, and hence are less liable to condensation in cold weather.

58 **A** = False **B** = True **C** = True **D** = False **E** = True

The accommodative effort needed to read a near object depends on the distance of the object, the power of the correcting lens and the distance of the correcting lens from the eye. This is because the correcting lens in different positions changes the vergence of rays of light differently for objects at varying distances from the eye. In general, accommodative effort is greater in myopia if contact lenses rather than spectacles are used. Conversely, accommodative effort is greater in hypermetropia if spectacles rather than contact lenses are used.

This is an important consideration when prescribing for a presbyopic subject. Prescription of contact lens to a myopic presbyopic subject may lead to difficulties in reading.

Accommodative effort for a myopic subject is higher with than without glasses. Conversely, it is lower with than without glasses in hypermetropia.

59 **A** = False **B** = True **C** = False **D** = True **E** = True

The aim of designing best form lenses is to eliminate troublesome optical aberrations. The most important aberration to eliminate is oblique astigmatism. Otherwise, the subject cannot see distinctly when the eyes move around. It has been found that there are two forms of lenses where the astigmatic effect is eliminated for a range of power (+7D to –22D). The Ostwald form has a shallow curvature, while the Wollaston form has a deep meniscus.

The base curve is the flatter curve, and the steeper surface is called the second surface. For a convex lens higher than 8D, no spherical lens can eliminate oblique astigmatism, and an aspherical lens is needed.

To reduce oblique astigmatism, a more concave surface should be placed next to the eye, and a more convex surface away from the eye. Hence, the base curve (flatter curve) of a positive meniscus is the back surface.

60 **A** = False **B** = True **C** = True **D** = True **E** = False

The keratometer measures the radius of curvature in two meridians of the anterior surface of the cornea on points about 1.5 mm from the centre. The astigmatisms of the posterior surface of the cornea and the lens are ignored. Hence, it does not measure the astigmatism of the whole eye, and is not an alternative to retinoscopy. When performed in conjunction with the axial length of the eye, it is useful in predicting the power of intra-ocular lenses required and for fitting contact lenses.

TEST 2
Questions

1 A spectrum can be produced by passing white light through

A a convex lens
B a concave lens
C a triangular prism
D a diffraction grating
E a rectangular block of glass

2 Light of wavelength 320 nm can

A penetrate the cornea
B penetrate the crystalline lens
C reach the retina in a phakic eye
D reach the retina in an aphakic eye
E be absorbed by the retinal pigment epithelium

3 Regarding diffraction of light

A the radius of the central bright spot (airy disc) decreases with increasing wavelength of light
B the radius of the central bright spot decreases with increasing distance from the aperture to the screen
C the radius of the central bright spot decreases with increasing diameter of the aperture
D the smaller the central bright spot, the greater is the resolving power of the optical system
E it is synonymous with scattering

4 The following statements regarding prisms is/are true.

A The emergent ray is deviated away from the base of prism.
B The angle of deviation *D* is the minimum when the angle of incidence equals the angle of emergence.
C The angle of deviation equals half of the refracting angle of the prism irrespective of the optical density of the prism.
D A prism of 1 Δ power produces an apparent displacement of 1 mm at 1 metre.
E One centrad (1 ∇) is slightly more than 1 Δ.

5 In the emmetropic eye

A the radius of curvature of the anterior surface of the cornea is about 8 mm
B the optical centre of the eye is in the middle of the crystalline lens
C the image of an object on the fovea is erect
D the crystalline lens contributes more than 50% of the refracting power of the eye
E a cataract in the nucleus of the lens affects vision more than a posterior subcapsular cataract

6 The following statements is/are true regarding a laser.

A It stands for light amplification stimulated emission of radiation.
B Its production requires 'optical pumping'.
C The length between the mirrors in the laser producing tubes must be equal to multiples of the wavelength of the light produced.
D The mirrors in the laser producing tube must be totally opaque.
E It may cause blinding retinal burn.

7 Regarding Sturm's conoid with greatest and least curvature in the horizontal and vertical meridian respectively

A the circle of least diffusion represents the point when the subject sees the object most clearly
B the focal interval of Sturm is the distance between the refracting surface and the focus for the vertical ray
C the circle of least diffusion always lies between the two foci for vertical and horizontal rays
D there may be more than one circle of least diffusion assuming that the vertical and horizontal foci are different
E it can be accurately represented in the plane of a piece of paper

8 In the conoid of Sturm produced by a lens system of +3DS/+2DC × 80°

A the interval of Sturm is 13.3 cm
B the circle of least diffusion is 25 cm from the lens
C addition of +1DS to the system will not alter the interval of Sturm
D addition of −0.5DS to the system will increase the interval of Sturm
E addition of −1DC × 80° will move the distal focal line further away

9 In the process of neutralisation of a lens with another lens

A the lenses should be held with their optical centres opposite each other
B they should be separated by at least 0.5 cm
C a prism is sometimes required
D the spherical/cylindrical component should be neutralised before the prismatic component
E a constant displacement without relative movement of the arms of the cross denotes a prismatic effect

10 Convexo-concave lenses

A are also called menisci
B always act as convex lenses
C act as a concave lens if the radius of curvature from the convex surface is longer than the radius of curvature of the concave surface
D have their optical centres midway between the two surfaces
E are often used as trial lenses

11 A myopic subject requires a spectacle correction of –10D. The back vertex distance is 15 mm.

A A contact lens of –10D will result in under-correction.
B A contact lens of –8.75D is needed.
C A contact lens of –11.25D will result in over-correction.
D A spectacle lens of –8.5D will result in under-correction if the back vertex distance is 10 mm.
E A spectacle lens of –10.5D will result in over-correction if the back vertex distance is 10 mm.

12 A subject has a far point of 2 m and a near point of 25 cm.

A The range of accommodation is 1.75 m.
B The amplitude of accommodation is 3.5D.
C Addition of +1D would increase the range of accommodation.
D Addition of +1D would decrease the amplitude of accommodation.
E Addition of –1D would increase the far point.

13 The following statements about myopia is/are true.

A The image is in front of the retina.
B It is always caused by an abnormally long eyeball.
C It can be corrected by a concave lens.
D It can be corrected by accommodation.
E The visual acuity is improved by screwing up the eye.

14 The following is/are true of hypermetropia.

A Latent hypermetropia is caused by the ciliary muscle tone.
B Total hypermetropia equals the sum of latent and absolute hypermetropia.
C Total hypermetropia equals the sum of absolute and facultative hypermetropia.
D Facultative hypermetropia is that part of hypermetropia that can be relaxed by accommodation.
E Manifest hypertropia increases with age.

15 The following statements regarding astigmatism is/are true.

- **A** Compound hypermetropic astigmatism may be corrected by a cylindrical lens alone.
- **B** Simple hypermetropic astigmatism may be corrected by a spherical lens.
- **C** Simple myopic astigmatism may be corrected by a cylindrical lens alone.
- **D** Compound myopic astigmatism may be corrected by a combination of concave and cylindrical lenses.
- **E** Mixed astigmatism may be corrected by a combination of spherical and cylindrical lenses.

16 Accommodation

- **A** is caused by changes in radius of curvature of the posterior surface of the lens
- **B** has a diverging effect on an incoming ray of light
- **C** puts the eye into static refraction
- **D** occurs in both eyes simultaneously
- **E** can be controlled both voluntarily and involuntarily

17 Index myopia

- **A** is caused by increased curvature of the cornea
- **B** may be caused by an abnormal refractive index of the lens
- **C** may precede senile cataracts
- **D** is the commonest cause of myopia
- **E** can be corrected by a concave lens

18 Presbyopia

- **A** is a normal aging process
- **B** is corrected by a concave lens
- **C** becomes manifest earlier in life in uncorrected hypermetropic eyes
- **D** requires the same correction in reading a book and in reading music
- **E** is associated with increasing the near point beyond reading distance

19 The amplitude of accommodation

- **A** is about 25D in infancy
- **B** is less than 1D in subjects aged over 65
- **C** can be determined by measuring the refractive error
- **D** can be determined by measuring the far and the near point
- **E** is about 6D at aged 40

20 Both surfaces of a biconvex lens have radii of curvature of 5 cm. Assuming that the refractive index of the lens is 1.5

A the refractive power at one surface is more than 12D
B the refractive power of the lens is less than a similar lens with radii of curvature of 4 cm for both surfaces
C the refractive power at one surface is the same if the lens is placed in water
D the refractive power of the lens is independent of the refractive index of the lens
E the refractive power of the lens is 20D

21 The following lenses are equivalent

A +5DS/–2DC × 20° and +3DS/–2DC × 110°
B –6DS/+3DC × 50° and –3DS/–3DC × 140°
C +8DS/+2DC × 45° and +10DS/–2DC × 45°
D –4DS/–3DC × 35° and –7DS/+3DC × 125°
E +9DS/+1DC × 10° and +9DS/–1DC × 100°

22 Problems in correction of aphakia by contact lenses include

A the inability of contact lenses to correct astigmatism
B infective corneal ulcers
C papillary conjunctivitis
D intolerance of contact lenses
E glaucoma

23 The back vertex power

A is synonymous with back vertex distance
B is the most important factor determining the power of a spectacle lens
C can be calculated from the posterior focal length of the lens
D is equal to the effective power if placed at the anterior focal plane of the eye
E is higher the longer the posterior focal length of the lens

24 The following statements about visual acuity is/are correct.

A The lower number of a Snellen's fraction is the distance between the subject and the test object at which the target subtends an angle of 1 minute at the retina.
B The Snellen's fraction is the same as the reciprocal of the visual angle subtended at the retina in minutes.
C A visual acuity of 6/9 is equivalent to a visual acuity of 9/6.
D The target of 6/18 is linearly three times larger than the target of 6/6.
E A visual acuity of 6/9 is equivalent to a visual acuity of 20/30.

25 The stenopaeic slit

A is used in conjunction with the slit lamp
B is useful in detecting regular astigmatism
C is useful in detecting myopia
D is useful in detecting irregular astigmatism
E allows only rays of lights in a particular meridian to enter the eye

26 Regarding measurement of visual acuity by Snellen chart

A the letter 'L' can be recognised more easily than 'K'
B the letter 'E' can be recognised more easily than 'B'
C the letter 'S' can be recognised more easily than 'A'
D single letters can be recognised more easily than a row of letters of the same size
E Landolt's rings may be used instead for illiterate persons

27 The following statements is/are true about image distortion.

A Convex lenses usually causes 'pin-cushion' distortion.
B Concave lenses usually causes 'barrel' distortion.
C It is related to the prismatic effect of the lens.
D It is independent of the power of the lens.
E It affects the peripheral more than the central field of vision.

28 A –2D spherical lens placed in front of a subject causes neutralisation at two different meridians at 90° when the examiner is 50 cm away from the subject on retinoscopy.

A A convex lens should be prescribed.
B The subject is –4D myopic.
C Neutralisation has been achieved for all meridians.
D A cylindical trial lens should be used.
E The subject has astigmatism.

29 The following statements about cholinergic blocking agents is/are true.

A They cause pupil dilatation.
B They cause paralysis of accommodation.
C Tropicamide has a better cycloplegic action than mydriatic action.
D The mydriatic effect of atropine can lasts up to 2 weeks.
E The cycloplegic action of cyclopentolate can last for 24 hours.

30 The streak retinsocope

A produces a circular source of light
B uses a plane mirror
C allows the band outside the pupils to be better visualised
D produces a more dramatic appearance at neutralisation
E may result in a more accurate estimation of the cylindrical axis than a non-streak retinoscope

31 In retinoscopy

A if the points of reversals are the same in all meridians, the refractive error is purely cylindrical
B if the points of reversals are different in two meridians at right angles to each other, there is astigmatism
C if the edge of the illuminated area of the retina appears oblique in the vertical meridian, the axis of the cylinder is oblique
D if there is 'with movement' in one meridian and 'against movement' in another when there is no lens in front of the subject's eye, there must be mixed astigmatism if the working distance is 2/3 m
E if a +1.5 lens is placed in front of the subject's eye, and there is 'with movement' in one meridian but 'against movement' in another, there must be mixed astigmatism if the working distance is 2/3 m

32 The following is/are true regarding performing retinoscopy in an eye with opacities in the optical media.

A The retinoscopy reflex may be abolished.
B Increasing the working distance may render the procedure easier.
C Using a brighter source of light may render the procedure easier.
D Cycloplegics may be used to render the procedure easier.
E The results obtained may be inaccurate.

33 The following statements about astigmatic dial is/are true.

A It is another name for astigmatic fan.
B It consists of a series of lines arranged radially every 45°.
C Accommodation should be prevented for the test to be effective.
D If two lines are seen to be more distinct than other lines, the axis of astigmatism is near to the centre of the two lines.
E A minus lens can be placed with its axis perpendicular to the darkest line to correct the astigmatism.

34 The field of view in direct ophthalmoscopy

A is smaller than that of indirect ophthalmoscopy
B is the total area of the retina that can be seen by manoeuvring the instrument in different directions
C is smaller in a myopic eye than a hypermetropic eye
D is independent of pupil size
E is larger the further the instrument is from the patient's eye

35 The following is/are true about the use of the indirect ophthalmoscope.

A The field of vision is larger in hypermetropia than myopia.
B Use of a +20D condensing lens gives a larger magnification than a +13D lens.
C The image size decreases as the condensing lens moves away from the subject's eye in myopia.
D The image size remains constant irrespective of the distance between the condensing lens and the subject's eye in emmetropia.
E It can give a stereoscopic view of the fundus.

36 The prismatic correction in glasses

A is occasionally useful for squints
B is limited by the variation of prismatic deviation with the tilt of the prism
C is helped by chromatic aberration
D may cause optical distortion
E may be placed horizontally, vertically or obliquely

37 Prisms can be used

A to assess the size of squints
B to test for microtropia
C to relieve diplopia in decompensated squints
D to distinguish between hysterical and true blindness
E in an applanation tonometer

38 The focimeter

A contains a standard lens
B should be used with the surface of the lens being tested facing the lens rest
C measures the refractive power of the whole lens
D contains a telescope system which, when focused, renders light entering it parallel
E gives a reading of the focal length of the lens being tested

39 A pachymeter can be used to measure

A the axial length of the eye
B the diameter of a crystalline lens
C corneal thickness
D anterior chamber depth
E the thickness of a contact lens *in situ*

40 Regarding the use of a slit lamp

A iris atrophy is best demonstrated by lateral illumination
B the anterior capsule of the lens can be examined by diffuse illumination
C the corneal endothelium is best examined by specular reflection
D lateral illumination is the most commonly used method
E the illuminating and the viewing systems of the slit lamp should be at 45° to each other in retro-illumination

41 Hirschberg test

A is useful in uncooperative subjects
B can be used to distinguish between pseudo-esotropia caused by epicanthic fold and convergent squint
C can be used to distinguish between pseudo-exotropia and divergent squint
D can be used to distinguish between squints and abnormal angle kappa
E should be followed by the cover tests

42 The following statements is/are true regarding prescription of bifocal lenses.

A The vertical muscle balance should be assessed.
B The horizontal muscle balance should be assessed.
C It is particularly useful for those who work at height.
D Subjects should move their heads upwards for near vision and downwards for distant vision to some degree.
E It is inadvisable to prescribe in full for old people who have never tried it before.

43 Iseikonic lenses

A are used to correct astigmatism
B are always effective in aniseikonia
C can be clipped on to the present spectacles for a trial period
D can be combined with lenses with refractive correction
E can be combined with bifocal lenses

44 The following is/are true for a telescope system containing two convex lenses.

A The final image is real.
B The final image is erect.
C The final image is laterally inverted.
D The final image is magnified.
E The distance between the lenses can be determined by the focal lengths of the two lenses.

45 The following is/are disadvantages of a hand magnifier over a stand magnifier.

A It is expensive.
B It leaves hands occupied.
C It has a lower magnification.
D It is bulky.
E It is less portable.

46 In the evaluation of the fitting of contact lenses

A instillation of fluorescein viewed with blue filter is useful
B a well-fitted lens should gravitate towards the bottom of the eye
C the lens should move a little with each blink
D there should be maximum fluorescein in the intermediate zone
E evaluation should be done at least 20 minutes after fitting of the lens

47 Therapeutic soft lenses are of value in

A bullous keratopathy
B chemical burns
C pseudomonas corneal infection
D recurrent corneal erosions
E trichiasis

48 The following are recognised complications of contact lens wear

A giant papillary conjunctivitis
B retinal detachment
C corneal oedema
D vascular pannus formation
E angle closure glaucoma

49 The water content of a soft contact lens

A can be measured by a refractometer
B is useful in identifying the material of the lens
C affects the refractive index of the lens
D is not related to the oxygen transmissibility of the lens
E is directly proportional to its refractive index

50 The following is/are liable to cause degradation of contact lens material over time.

A Heat.
B Ultraviolet radiation.
C Moisture.
D Materials containing impurities.
E Repeated stress.

51 Spherical aberration of the eye

A is increased by a pinhole aperture
B contributes to 'night myopia'
C is reduced by the aspheric surface of the cornea
D affects the central field more than the peripheral field
E increases with dilatation of the pupil

52 The relative spectacle magnification in axial ametropia

A when corrected by a lens placed at the anterior focal plane of the eye, is always 1
B when corrected by a lens placed at the anterior focal plane of the eye, is governed by Knapp's Rule
C when corrected by a lens placed behind the anterior focal plane of the eye, is less than 1 in myopia
D when corrected by a lens placed in front of the anterior focal plane of the eye, is more than 1 in myopia
E is the same as in refractive ametropia

53 The following statements about relative accommodation is/are true.

A The positive portion can be measured by the highest negative lens that can be placed in front of an eye without blurring.
B The negative portion can be measured by the highest positive lens that can be placed in front of an eye without blurring.
C The positive portion should be kept as large as possible for comfort.
D The positive portion increases as the object moves away from the eye.
E The negative portion increases as the object moves towards the eye.

54 A spectacle lens of +5DS/+5DC at 90° is decentred.

A Decentration of 0.5 cm horizontally would give a prismatic effect of 5 Δ.
B Decentration of 1 cm vertically would give a prismatic effect of 5 Δ.
C Decentration of 1 mm horizontally would give a prismatic effect of 1 Δ.
D Decentration would give rise to the same prismatic effect irrespective of the direction.
E The prismatic effect at 30° can be calculated from the above.

55 The vergence power (fusional reserves)

A is the measure of the ability of the eyes to maintain binocular single vision when forced into vergence
B can be estimated by using a prism-bar and testing for the onset of diplopia
C is best measured using a major amblyoscope
D is different for vertical and horizontal vergence
E is the same for divergence and convergence

56 The following about binocular tests is/are true.

A They should be performed at the beginning of subjective refraction.
B Binocular visual acuity is better than each eye alone if there is significant difference between the visual acuities in each eye.
C They may be carried out by dissociation with vertical prisms.
D They may be carried out comparing the results of the Duochrome test in each eye while the eyes are dissociated.
E They may include a test of muscle imbalance.

57 The following is/are true regarding the interpupillary distance.

A Determination for distance vision only is required.
B It is shorter for females than males.
C Facial asymmetry may cause inequalities of the two halves of the inter-pupillary distance.
D Inaccuracies in measurement may result in prismatic effects.
E It differs amongst individuals.

58 An aphakic subject requires a spectacle correction of +12D. The back vertex distance is 15 mm, and the anterior chamber is 5 mm deep.

A A contact lens of 10.75D results in under-correction.
B A contact lens of 12.75D results in over-correction.
C A contact lens of 16D results in over-correction.
D A posterior chamber intra-ocular lens of 14.75D results in under-correction.
E A posterior chamber intra-ocular lens of 15.75D is required.

59 The following about best form lenses for astigmatic lenses is/are true.

A They are simpler to produce than those for spherical lenses.
B They are produced mainly by bending the lens.
C The ideal lenses with no aberration are often unachievable.
D They often suffer from troublesome reflection from bent surfaces.
E Computers may assist in the design.

60 Abnormalities in accommodation

A typically cause blurring of vision
B may cause transient symptoms
C usually affect *either* near *or* distance vision only
D include presbyopia
E may cause headaches

TEST 2
Answers

1 **A** = False **B** = False **C** = True **D** = True **E** = False

A spectrum is produced by the separation of white light into its components. It depends on the fact that the refractive index of glass differs slightly for different wavelengths of light. When white light is passed through a concave lens, convex lens and rectangular block of glass, different wavelengths of light are separated inside the glass, but they are recombined when emerging from the glass. Spectrum can also be produced by a diffraction grating, although the order of the colours is reversed (see Fig. 2.1).

2 **A** = True **B** = False **C** = False **D** = True **E** = True

Light in the visible spectrum penetrates through all the refractive media in the eye. The cornea allows light in the ultraviolet region to penetrate, but absorbs light below about 295 nm. The crystalline lens absorbs light of wavelength shorter than 350 nm. This protects the retina against the ultraviolet light, which may cause pathological changes. Light of all wavelengths is well absorbed by both the pigment epithelium at the back of the iris and the pigment retinal epithelium.

3 **A** = False **B** = False **C** = True **D** = True **E** = False

The diffraction pattern from a circular aperture of a point source of light is a central bright spot (airy disc) surrounded by concentric less bright fringes of light resulting from diffraction. The radius of the airy disc (image) is given by the formula

$$\frac{1.22 \times \text{wavelength of light} \times \text{distance of aperture to image}}{\text{diameter of aperture}}$$

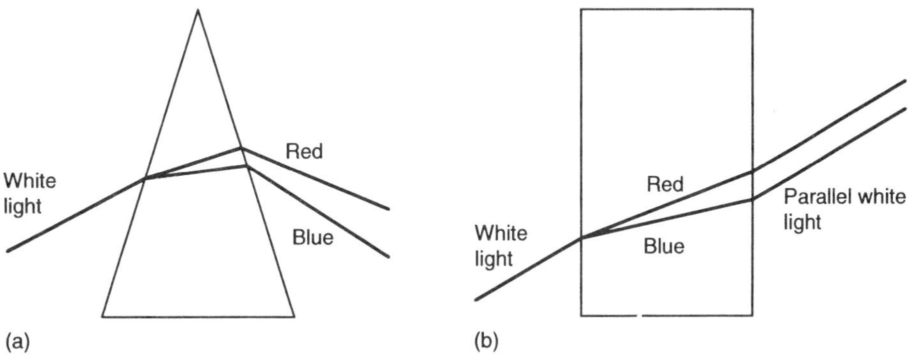

Figure 2.1 (a) Production of a spectrum by a prism; and (b) passage of light of different colours through a glass block.

An increase in the wavelength of light or the distance of the aperture from image, and reduced diameter aperture would increase the radius of the airy disc. The smaller the diameter of the airy disc, the higher is the resolution power. Scattering refers to the phenomenon of light reflected off small particles when travelling through a medium, and is different from diffraction.

4 **A** = False **B** = True **C** = False **D** = False **E** = True

A prism is formed by two plane surfaces of a refractive medium meeting at an angle, the **apical** or **refracting angle**. The opposite surface to the apical angle is called the **base** of the prism. The axis of the prism is defined by the line bisecting the apical angle. Light is deviated towards the base of the prism. **The angle of deviation** is the angle between the incidence and emergence ray of light. This is a **minimum** when the angle of incidence and the angle of emergence are equal, that is when the light passes symmetrically through the prism. In crown glass (refractive index 1.5), this angle is approximately half the apical angle (see Fig. 2.2(a)).

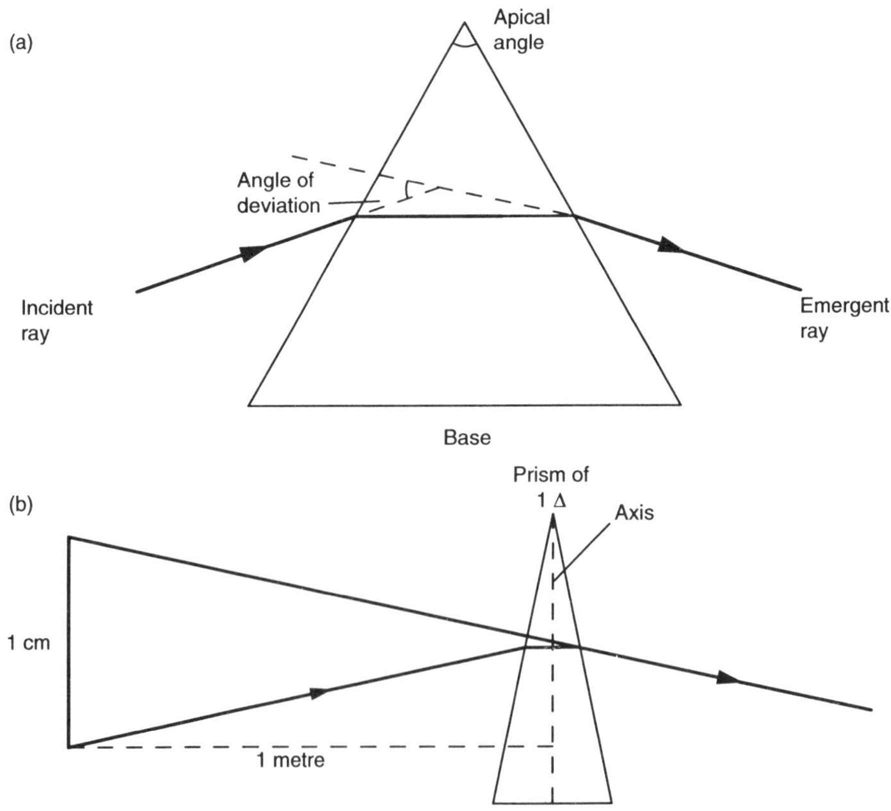

Figure 2.2 (a) The triangular prism; and (b) definition of prism-dioptre.

The power of the prism can be measured by:

a. the apical angle.

b. the angle of deviation when the incidence and emergence rays are symmetrical.

c. the **Prism dioptre** (Δ). A prism of 1 Δ produces a linear displacement of 1 cm of an objected situated at a distance of 1 m (see Fig. 2.2(b)).

This is the most commonly used measurement.

The centrad differs from the prism dioptre only in that the image displacement is measured along an arc 1 m from the axis of the prism rather than on a vertical meridian. Its prismatic power is slightly more than a prism dioptre.

5 **A** = True **B** = False **C** = False **D** = False **E** = False

Also see Test 3 Question 5. The anterior surface of the cornea is nearly spherical, and the radius of curvature is about 8 mm. The centres of curvatures of the cornea and the anterior and posterior surfaces of the lens are all in the same line, and form the optical axis. The image at the retina is inverted, but the occipital cortex helps us the interpret the image correctly.

6 **A** = True **B** = True **C** = True **D** = False **E** = True

Laser stands for Light Amplification by the Stimulated Emission of Radiation. Normally, there is an ongoing absorption of photons of correct frequency by atoms which causes a transition of particles to a higher energy state. There is also spontaneous emissions of photons when the atoms fall from the higher to lower energy state. In stimulated emission, an incoming photon of appropriate frequency causes the atom to 'resonate' and induces atoms to fall from higher to lower energy states, so that there are two outgoing photons of the same frequency. The process is then repeated.

A laser can be produced in a gas tube with the ends coated with aluminium to act as 'mirrors'. The 'mirrors' are exactly parallel to each other, and are separated by multiples of the wavelength of the light produced. Initially, the stimulated photons are emitted in all directions. However, only those that travel parallel to the tube increase. One 'mirror' is made slightly transparent so that light can escape as a laser beam.

The beam intensity is very high, so that it can cause retinal burn.

7 **A** = True **B** = False **C** = True **D** = False **E** = False

See Fig. 2.3.

Sturm's conoid is a ray diagram representation of a regular astigmatic refracting surface (such as the astigmatic eye) with the horizontal and vertical meridians having different foci. The distance between the horizontal and vertical foci is known as the interval of Sturm. The circle of least diffusion (C) occurs when the horizontal rays are diverging as much as the vertical rays are converging. Hence, it is always between the two foci, and there is only one circle of least diffusion. The circle of least diffusion represents the point at which the subject sees the object most clearly.

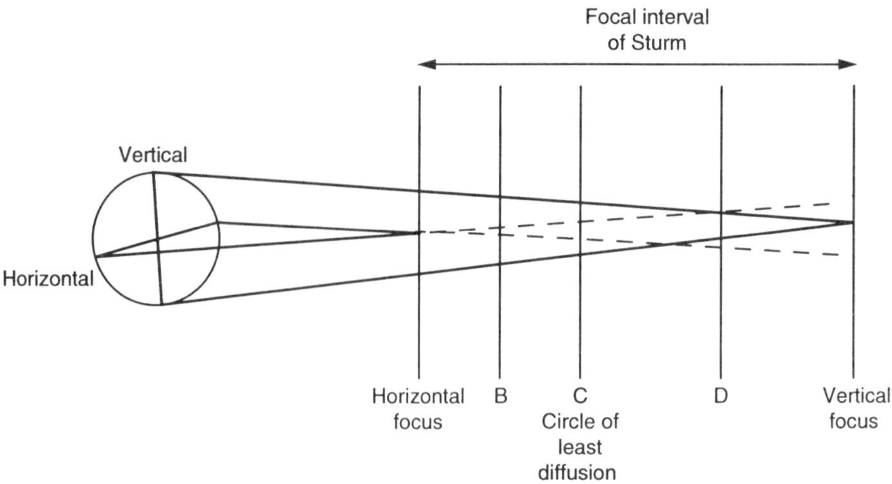

Figure 2.3 The conoid of Sturm.

8 **A** = True **B** = True **C** = False **D** = True **E** = False

See also Test 2 Question 7. The interval of Sturm is the distance between the two focal lines. In a lens system of +3DS/+2DC × 80°, the two focal lines have dioptrical power of 3D and 5D. Hence, they are 1/3 m = 33.3 cm and 1/5 m = 20 cm away from the lens. The interval of Sturm is 33.3 – 20 = 13.3 cm.

The circle of least diffusion is midway dioptrically between the two focal lines. Hence, the power at the circle of least diffusion is (3 + 5)/2 = 4D. It is 1/4 = 0.25 m = 25 cm away from the lens.

Addition of 1 DS to the system will result in a lens system of +4DS/+2 DC × 80°. The power at the focal points are 4D and 6D. Hence, they are 1/4 = 0.25 m and 1/6 = 0.167 m away from the lens. The interval of Sturm is 25 – 16.7 cm = 8.3 cm.

Similarly, addition of –0.5 DS to the system will result in the powers at the focal lines being 2.5D and 4.5D. The focal lines are 1/2.5 m = 40 cm and 1/4.5 m = 22.2 cm away from the lens. The interval of Sturm is (40 – 22.2) cm = 17.8 cm.

Addition of –1DC × 80° will result in a lens system of +3DS/+1DC × 80°. The distal focal line has a power of 4D, and hence is 25 cm away.

9 **A** = True **B** = False **C** = True **D** = True **E** = True

In the neutralisation of the lenses, the lenses should be held close together with their optical centres opposite each other. Otherwise, the convex lens would require a stronger concave lens to neutralise. The spherical and cylindrical components should be neutralised before the prismatic component. A positive spherical lens causes 'against movement', while a negative spherical lens causes 'with movement'. A prismatic lens causes constant displacement without relative movement of the cross.

10 **A** = True **B** = False **C** = True **D** = False **E** = False

A convexo-concave lens has a concave surface on one side and convex surface on the other. It is also called a meniscus. Its total refractive power is equal to the sum of the refractive power of the two surfaces. Hence, it can either act as a concave or convex lens depending on which surface has the greater refractive power. If the radius of curvature of the convex surface is longer than that at the concave surface, the convex surface has a less refractive power than the concave surface. Hence the lens acts as a concave lens. The optical centre is outside the lens.

Trial lenses are usually plano-convex or plano-concave lenses, with the convex or concave surface nearest to the eye.

11 **A** = False **B** = True **C** = True **D** = True **E** = True

Applying the formula in Test 1 Question 11 that

$$F_d = \frac{F}{1 + dF}$$

where F_d = refractive power of the correcting lens placed d metres away from the eye (negative if moved towards the eye), F = refractive power of the correcting lens of original lens in dioptres (+ve for convex lens, –ve for concave lens) and d = distance moved away from eye (negative if moved towards eye).

If a contact lens is used

$$\begin{aligned} F &= -10\text{D} \\ d &= -0.015\text{ m} \\ F_d &= -10/(1+ (-0.015 \times -10)) \\ &= -8.7\text{D} \end{aligned}$$

If the back vertex distance is 10 mm

$$\begin{aligned} F &= -10\text{D} \\ d &= -0.005\text{ m} \\ F_d &= -10/(1 + (-0.005 \times -10)) \\ &= -9.52\text{D} \end{aligned}$$

12 **A** = True **B** = True **C** = False **D** = False **E** = True

A subject with a far point of 2 m is 1/2 = 0.5D myopic. Focusing on an object at 25 cm requires a refractive power of 1/0.25 = 4D. Hence, the amplitude of accommodation is (4 – 0.5) = 3.5D.

The range of accommodation is the difference between the distance of the far point and the near point = 2 m – 25 cm = 1.75 m.

Addition of +1D would render the subject 1.5D myopic. The far point is (1/1.5) = 67 cm. The near point is (1/(1.5 + 3.5)) m = 20 cm. Hence the range of accommodation is (67 – 20) = 47 cm. The amplitude of accommodation remains the same irrespective of the lens used.

Addition of −1D lens renders the subject 0.5D hypermetropic. Hence, the far point is at infinity (when accommodation of 0.5D is used). The range of accommodation is increased.

13 **A** = True **B** = False **C** = True **D** = False **E** = True

Also see Test 1 Question 13. Myopia can be caused by either an abnormally long eyeball (axial myopia) or excessive refractive power of the eye (refractive myopia). Excessive refractive power of the eye can be due either to increased curvature of the cornea or increased refractive index of the lens. The image is formed in front of the retina. It can be corrected by a diverging (concave) lens. One can see more clearly by screwing up the eyes, due to the pinhole effect.

14 **A** = True **B** = False **C** = False **D** = True **E** = True

Also see Test 1 Question 14. Total hypermetropia equals the sum of latent and manifest hypermetropia. Manifest hypermetropia equals the sum of absolute and facultative hypermetropia. Manifest hypertropia increases with age, as the amplitude of accommodation decreases.

15 **A** = False **B** = False **C** = True **D** = True **E** = True

Compound hypermetropic astigmatism occurs when the eye is hypermetropic for both vertical and horizontal meridians, and requires correction with a combination of a positive spherical (convex) lens and a positive cylindrical lens. Simple hypermetropic astigmatism occurs when the focus for one meridian is on the retina, but the eye is hypermetropic for the other meridian. A positive cylindrical lens is required for correction. Simple myopic astigmatism similarly requires correction with a negative cylindrical lens. Mixed astigmatism occurs when the retina falls between the two foci, and requires a combination of a spherical lens and a cylindrical lens of the opposite sign for correction.

16 **A** = False **B** = False **C** = False **D** = True **E** = True

See Test 1 Question 18.

17 **A** = False **B** = True **C** = True **D** = False **E** = True

Index myopia is caused by abnormal refractive indices of the media. This can be caused by an abnormally high refractive index in the cornea, aqueous, or the nucleus of the lens. Alternatively, it may be caused by an abnormally low refractive index in the vitreous (reducing the difference of refractive indices between the lens and the vitreous). An example of index myopia is nucleus sclerosis prior to the development of a nuclear cataract.

18 **A** = True **B** = False **C** = True **D** = False **E** = True

Presbyopia is caused by the physiological decline in the amplitude of accommodation with age. In newborn babies, the amplitude of accommodation is 14D. It is reduced to about 8D at 30 years old, 6D at 40 years, and less than 1D at 60 years.

Accommodation of 4D is required for a reading distance of 25 cm, and 3D is required for a reading distance of 33 cm. A convex lens is required to bring nearby objects into focus on the retina. Reading music is performed with the music score further away from the eye than a book, and hence requires less presbyopic correction.

As the amplitude of accommodation decreases, the near point of the eye moves further away than the distance required for comfortable reading, and blurring of vision occurs. Generally speaking, for comfortable reading, one third of the available amplitude of accommodation must be kept in reserve. In an uncorrected hypermetropic eye, part of the accommodation is used to overcome the hypermetropia. Hence, the hypermetropia manifests itself earlier in life, although the rate of physiological changes of the lens is the same. Conversely, in a myopic eye, presbyopia manifests itself later.

19 **A** = False **B** = True **C** = False **D** = True **E** = True

See also Test 4 Question 12 and Test 3 Question 19. The amplitude of accommodation in infancy is about 14D. The amplitude of accommodation cannot be determined from the refractive error (i.e. refraction when unaccommodated) alone, as the amplitude of accommodation is the difference of refraction while accommodated and while unaccommodated. It is possible to calculate the amplitude of accommodation from the far point and the near point:

Amplitude of accommodation = 1/[near point (in m)] – 1/[far point (in m)]

20 **A** = False **B** = False **C** = False **D** = False **E** = True

Applying the formula

$$\text{Power of surface (in dioptres)} = (n_1 - n_2)/r$$

where n_1 = refractive index of lens, n_2 = refractive index of the medium (air) and r = radius of curvature of lens in m (+ve if convex, –ve if concave)

Refractive power at one surface = (1.5 – 1)/0.05 = 10D.
Refractive power of the lens = 10 + 10D = 20D.

The refractive power of the lens is less if placed in water. The power of the each surface of the lens increases as the radii of curvature decreases.

21 **A** = False **B** = True **C** = False **D** = True **E** = False

The general formula for transposing lenses

X DS/*Y* DC at axis *r* degress

is (*X* + *Y*) DS/–*Y* DC at axis (*r* + 90)° if *r* is less than 90°, or (*X* + *Y*) DS/–*Y* DC at axis (*r* – 90)° if *r* is more than 90°.

In **A**, the transposition of +5DS/−2DC × 20° is +3DS/+2DC × 110°
In **C**, the transposition of +8DS/+2DC × 45° is +10DS/−2DC × 135°
In **E**, the transposition of +9DS/+1DC × 10° is +10DS/−1DC × 100°

22 **A** = False **B** = True **C** = True **D** = True **E** = False

Problems in correction of aphakia by contact lenses include intolerance of foreign bodies feeling, oedema of corneal epithelium from prolonged wear, papillary conjunctivitis, peripheral corneal vascularisation, recurrent corneal erosions and infective corneal ulcers. Pupil block glaucoma is a complication of intra-ocular lens implantation.

23 **A** = False **B** = True **C** = True **D** = True **E** = False

The back vertex power equals the refractive power of the back surface of a spectacle lens, and can be calculated from (1/posterior focal length in m). The effective power of the lens depends mainly on the back vertex power, and the distance between the cornea and the back surface of the lens. If it is placed at the anterior focal plane of the eye, the effective power equals the back vertex power. The back vertex distance is the distance between the back surface of the lens and the cornea.

24 **A** = True **B** = True **C** = False **D** = True **E** = True

The upper number of a Snellen's fraction is the distance between the test object and the subject, while the lower number is the distance from the eye at which the target would subtend an angle of 1 minute on the retina. Hence, the Snellen's fraction is the same as the reciprocal of the visual angle subtended at the retina in minutes. A visual acuity of 6/9 is lower than that of 9/6. The Snellen fraction is sometimes expressed in feet (especially in the USA), and an acuity of 6/9 (in metres) is equivalent to that of 20/30 (in feet), as the fraction is the same.

25 **A** = False **B** = True **C** = False **D** = False **E** = True

The stenopaeic slit consists of a slit with the width of a pinhole cut out from an opaque disc and mounted on a trial frame. It allows only rays of light in a particular meridian to enter the eye. It is therefore useful in detecting regular astigmatism when the spherical component of the refractive error is corrected. The cylindrical axis can be found by rotating the slit until maximum visual acuity is obtained. A spherical lens can be added to find the refractive power in that meridian, and the slit can be rotated 90° to determine refractive power in the other meridian. It is useful when other methods, such as retinoscopy or the cross-cylinder, are unsatisfactory.

26 **A** = True **B** = True **C** = False **D** = True **E** = True

It has been found that certain letters (e.g. L, A, J, E) are easier to recognise than others (e.g. Y, V, S, R, Q). Letters C, G, O, Q are difficult as they can be easily confused with one another. Hence, in some charts, the absolute size of letters in a single line of the chart had been changed slightly to adjust for the different legibility of the letters.

For illiterate persons, Landolt's rings (where there is a break in the ring either at the top, bottom, left or right), or recognition of the orientation of the letter E can be used.

27 **A** = True **B** = True **C** = True **D** = False **E** = True

Image distortion is caused by the increased prismatic effect of spectacle lenses at the periphery, causing distortion of the shape of the edges of objects. When convex lenses are used, the most peripheral regions (i.e. the corners) are more 'magnified' and hence cause the 'pin-cushion' distortion. When concave lenses are used, the most peripheral regions are 'magnified' least and hence cause the 'barrel' distortion. The higher the power of the lens, the more the distortion.

28 **A** = False **B** = True **C** = True **D** = False **E** = False

The refractive error of the subject can be determined by subtracting working distance from the lens used (lens used – working distance refractive power).

In the example, the working distance is 50 cm. Hence, the equivalent refractive power is (1/0.5) = 2D

The subject is (–2) – 2 = –4D (4D myopic). There is no astigmatism as the neutralisation at the two meridians are the same.

If neutralisation points are reached for two meridians, neutralisation would have occurred for all meridians.

29 **A** = True **B** = True **C** = False **D** = True **E** = True

Cholinergic blocking agents result in both pupil dilatation (mydriatic effect) and paralysis of accommodation (cycloplegic effect). Atropine is the strongest cycloplegic agent available, and mydriasis can last up to 2 weeks. Cyclopentolate is often used in children. Its effect starts from 30 minutes, and can last for between 2 to 24 hours. Tropicamide is the most short-acting. Its mydriatic effect is much stronger than its cycloplegic effect, and usually only lasts for a few hours.

30 **A** = False **B** = False **C** = True **D** = True **E** = True

A streak retinoscope uses a plano-cylindrical mirror to produce a source of light in the shape of a streak. It has advantages over the ordinary retinoscope in that the band of light outside the pupil can be more easily seen, and any misalignment between the reflex in and outside the pupil can be more easily discerned. Hence, the astigmatic axis can be determined more easily. Also, a more dramatic appearance occurs at neutralisation when the whole pupil is filled with light.

31 **A** = False **B** = True **C** = True **D** = False **E** = True

If the points of neutralisation are the same in all meridians, the refractive error is purely spherical. If they are different, there must be a cylindrical component. If the edge of the illuminated area of the retina appears oblique in the vertical meridian, the axis of the cylinder must be oblique.

In **D**, there is 'with movement' in one meridian but 'against movement' in another when no lenses are placed in front of the subject's eye at a working distance of 67 cm. The working distance is equivalent to (1/0.67) = 1.5D of myopia. Hence, it can be deduced that the refractive power in one meridian is more than 1.5D myopia and less than 1.5D myopia in the other. Hence, it is not conclusive that there must be mixed astigmatism.

In **E**, the working distance is equivalent to 1.5D myopia and a lens of +1.5D is placed in front of the eye. It can be concluded that one meridian must be more than (1.5 – 1.5) = 0D and the other less than 0D. Hence, there must be mixed astigmatism.

32 **A** = True **B** = False **C** = True **D** = True **E** = True

With opacities in the optical media, the retinoscopy reflex may be reduced or abolished. Retinoscopy may be made easier by reducing the working distance, using a brighter source of light, or dilating the pupil. The results obtained may be inaccurate.

33 **A** = False **B** = False **C** = True **D** = True **E** = True

An astigmatic dial is a subjective test to determine both the axis and the power of astigmatism. It consists of a series of dark lines radiating outwards radially every 30° for the whole of 360°. The principle is similar to that of the astigmatic fan, except that, for the astigmatic fan, the series of dark lines is arranged in a semi-circle. For the test to be effective, accommodation should be discouraged by the use of fogging by about +0.5D. The subject is asked which line on the dial appears darkest and clearest. If there is no difference amongst the lines, there is unlikely to be any astigmatism. If there is a group of lines which are darker than the others, the axis of astigmatism is likely to be in the centre of that group. In order to determine the power of the astigmatism, a minus cylinder of increasing power is placed with its axis at right angles to the darkest line, until all the lines appear equally dark. Positive cylinders are not used in this case, as the image is already in front of the retina. If positive cylinders are used, a negative spherical lens should be added at the same time.

34 **A** = True **B** = False **C** = True **D** = False **E** = False

The field of view is the area of the retina that can be seen at any one time with the instrument. A larger area of the retina can be seen with an indirect compared with the direct ophthalmoscope. The area of the retina seen is larger when the pupil is larger. Getting closer to the subject's eye also increases the area of retina seen. The myopic fundus is more enlarged compared with the hypermetropic fundus with the direct ophthalmoscope, but the field of view is smaller.

35 **A** = False **B** = False **C** = False **D** = True **E** = True

In indirect ophthalmoscopy, the field of vision is largest in myopia and smallest in hypermetropia, unlike direct ophthalmoscopy (see Fig. 2.4).

In emmetropia, the image is always formed at the focal point of the condensing lens, and hence remains constant. In myopia, the image size decreases as the condensing lens is moved towards the subject's eye, while in hypermetropia, the image size increases as the condensing lens is moved towards the subject's eye. This magnification equals the ratio of the focal length of the condensing lens to the distance between the nodal point of the eye and the retina. Hence, a less powerful condensing lens would give a higher magnification.

36 **A** = True **B** = True **C** = False **D** = True **E** = True

The prismatic correction in glasses is only very occasionally used due to chromatic aberration, optical distortion and the variation of prismatic power with tilt of the prism. It is sometimes used for squints not correctable by surgery. The prism may be placed horizontally, vertically or obliquely depending on the type of ocular motility problems.

37 **A** = True **B** = True **C** = True **D** = True **E** = True

Prisms can be used for diagnosis, treatment and in optical instruments.

For diagnostic purposes, prisms can be used to assess the size of squints by the **prism cover test** (neutralisation of squint in cover tests with prism bar). Similarly, phorias detected by the Maddox rod may be neutralised by prisms, and the degree of phorias measured. **Microtropia** may be measured by the four dioptre prism base-out test. **Hysterical blindness** can be detected as the 'blind' eye will move to fixate when a prism is placed in front of the 'seeing' eye.

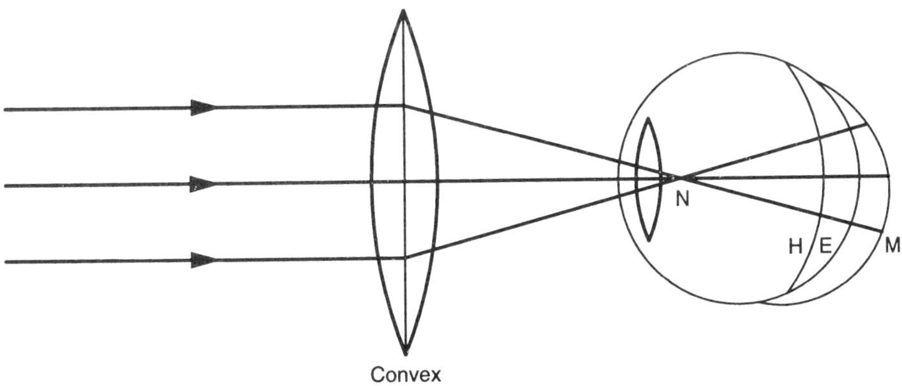

Figure 2.4 Indirect ophthalmoscopy: the field of vision is larger for the myopic eye and smaller for the hypermetropic eye. N = nodal point, H = hypermetropic retina, E = emmetropic point, M = myopic retina.

Prisms can be used in treatment of building up fusional reserves in patients with **convergence insufficiency**. During the exercise periods, **base-out prisms** are used. Prisms can also be used to **relieve diploplia** due to paralytic or non-paralytic squints in which surgery is not indicated. They can be either clipped on to the spectacles as temporary wear, as in **Fresnel prisms**, or as permanent wear by decentralisation of an existing spherical lens, or by mounting prisms in the spectacles.

Prisms can be used in optical instruments as in an applanation tonometer, slit lamp, or keratometer. They can also be used as very effective mirrors (by the principles of total internal reflection) to deviate light beams at certain angles, and to invert images. They can also be used to shorten the length of instruments (e.g. binoculars).

38 **A** = True **B** = True **C** = False **D** = True **E** = True

See Test 1 Question 38.

39 **A** = False **B** = False **C** = True **D** = True **E** = True

The pachymeter is used in conjunction with a slit lamp microscope. It utilises images formed from the anterior lens surface (image I), posterior corneal surface (image II), and the anterior lens surface (image III). Most instruments make use of the doubled image method—the examiner lines up the images of two surfaces, and the distance between the structures can be read off. Hence, the corneal thickness (from images I and II), and the anterior chamber depth (from images II and III) can be determined. The thickness of the contact lens *in situ* can also be determined from the image from reflection of the contact lens.

40 **A** = False **B** = True **C** = True **D** = False **E** = False

Direct focal illumination is the most common method of examination with the slit lamp, with the slit beam directed to the object under examination. The anterior capsule of the lens can be examined by **diffuse illumination**, when the slit beam is deliberately directed out of focus on the object under examination, so that the area is diffusely illuminated. The corneal endothelium is best examined by **specular reflection**, when the light reflected from the corneal endothelium is examined. **Retro-illumination** is the method of examining a structure of the eye by light reflected from the structure behind it. An example is examination of iris atrophy when light is reflected from the back of the eye to reveal atrophy of the iris. In retro-illumination, the illuminating and the viewing systems of the slit lamp should be in line with each other.

41 **A** = True **B** = True **C** = True **D** = False **E** = True

The Hirschberg test involves observation of the position of the corneal reflex in both eyes to determine whether a squint is present. It is a rough subjective method of estimating the size of squints. In pseudo-squints caused by the appearance of facial features, the position of the corneal reflex is central, and can be distinguished from true squints. In abnormal angle kappa (where the angle between the optical axis and the anatomical axis is abnormal) the corneal reflex would not be central, and can simulate squints.

42 **A** = True **B** = True **C** = False **D** = True **E** = True

When bifocal lenses are prescribed, horizontal and vertical muscle balance for both near and distant vision must be assessed, as prismatic effects of bifocal lenses can be significant. Bifocal lenses are unsuitable for those who are particularly liable to fall, or those who suffer from vertigo. They should not be given to those working with machinery or at height for safety reasons. Subjects wearing bifocal lenses have to learn, to some degree, to move their heads upwards for near vision and downwards for distant vision, which is the opposite to what they are accustomed to doing.

43 **A** = False **B** = False **C** = True **D** = True **E** = True

Iseikonic lenses are afocal lenses with magnification but no overall refractive power.They are used to correct aniseikonia (difference in image magnification between the two eyes), but are not always effective. As they are expensive to make, they are sometimes clipped on to the present spectacles for a trial period. They can be combined with lenses for correction of refractive error, or with bifocal lenses.

44 **A** = False **B** = False **C** = True **D** = True **E** = True

Also see Test 4 Question 45. In an astronomical telescope, the first lens produces an inverted real image, and the second lens acts as a magnifier to produce an inverted virtual magnified image. The distance between the lens equals the sum of the focal lengths of the two convex lenses.

45 **A** = False **B** = True **C** = False **D** = False **E** = False

Hand magnifiers are generally cheap, portable and can be immediately available. Magnification can be up to 15 times. However, hands are occupied when in use. Stand magnifiers leave hands free. However, they are bulky and generally have a lower magnification (up to 8 times) than hand magnifiers.

46 **A** = True **B** = False **C** = True **D** = False **E** = True

Evaluation of the trial lens should take place 20 minutes after insertion at the earliest, to allow time for the eye to get used to the trial lens and for reflex tearing to cease. The fit should be best evaluated using fluorescein viewed with a blue filter with the slit lamp. The lens should be well centred, move slightly with movement of the eye and blinking, and return to its original position. There should be maximum fluorescein in the centre, diminishing in the intermediate zone, and increasing again in the periphery.

47 **A** = True **B** = True **C** = False **D** = True **E** = True

Soft contact lenses are useful in treating a number of pathological conditions due to their flexibility and permeability. Hence, they are useful in treating bullous keratopathy, chemical burns, recurrent corneal erosions and trichiasis where hard corneal lenses may abrade the cornea further. Pseudomonas corneal infection is a serious infection requiring intensive topical and subconjunctival antibiotics. Contact lenses should not be used.

48 **A** = True **B** = False **C** = True **D** = True **E** = False

Complications of contact lens wear include punctate keratitis, corneal inflammatory reaction with corneal oedema, neovascularisation in the cornea, corneal infection, blepharitis and follicular/papillary conjunctivitis. Giant papillary conjunctivitis (with severe papillary reaction, mucous discharge, lens-lid adhesion) is particularly characteristic of contact lens wear.

49 **A** = True **B** = True **C** = True **D** = False **E** = False

The refractive index of a contact lens decreases as the water content increases. It is therefore useful in identifying the material of the contact lens. The water content can be measured by a refractometer. As water content increases, the oxygen transmissibility increases for a given lens thickness.

50 **A** = True **B** = True **C** = False **D** = True **E** = True

Factors likely to cause degradation of contact lens material over time include heat, ultraviolet radiation, chemicals, impurities and repeated stress.

51 **A** = False **B** = True **C** = True **D** = False **E** = True

Spherical aberration is caused by greater refraction at the periphery of the lens than the centre of the lens. It affects the periphery field more than the central field of vision. Hence, spherical aberration of the eye is reduced by a pinhole aperture and increased by dilatation of the pupil. Spherical aberration of the eye is reduced by the aspheric surface of the cornea—the periphery is less curved than the centre. It is also reduced by the refractive index of the nucleus of the crystalline lens being higher than the periphery of the lens.

'Night myopia' is caused by two factors. The first factor is that the rods in the dark-adapted retina are more sensitive to blue than red light. The second factor is that the pupil is more dilated, and that the peripheral rays of light become more important. Since the peripheral rays of light are refracted more than those in the centre, the image is focused in front of the retina, rendering the subject myopic.

52 **A** = True **B** = True **C** = False **D** = False **E** = False

See Test 1 question 52.

53 **A** = True **B** = True **C** = True **D** = True **E** = True

See also Test 1 Question 53. The importance of relative accommodation is that the positive portion should be larger than the negative portion for comfort. It has been estimated that at least one third of accommodation should be kept in reserve.

54 **A** = True **B** = True **C** = True **D** = False **E** = True

In the lens system of +5DS/+5DC at 90°, the power is 10D at 0° and 5D at 90°.
In **A**, prismatic effect = $0.5 \times 10 = 5\ \Delta$.
In **B**, prismatic effect = $1 \times 5 = 5\ \Delta$.
In **C**, prismatic effect = $0.1 \times 10 = 1\ \Delta$.
The prismatic effect would be greater in the horizontal than the vertical direction.

The prismatic effect by decentration of 1 cm at 30° to the horizontal can be calculated either graphically or by trigonometry/Pythagoras' Theorem.

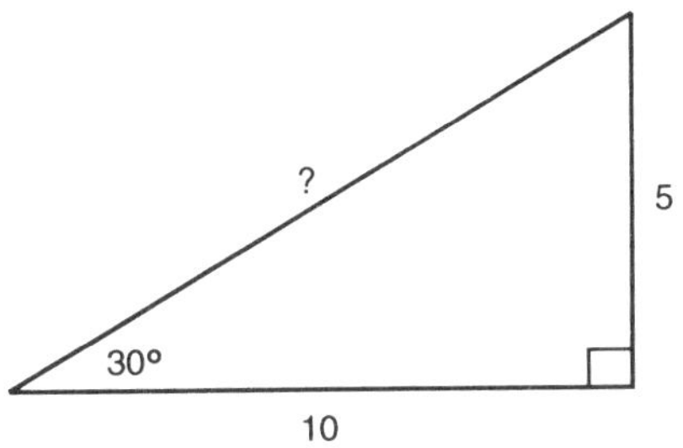

55 **A** = True **B** = True **C** = True **D** = True **E** = False

The fusional reserves are a measure of the ability of the eyes to maintain binocular single vision when forced to vergence (including convergence, divergence, or vertical vergence). They can be simply measured with graded prisms (as in a prism bar) and they measure the maximum prism lens that can be added before diplopia occurs. Alternatively, they can be more accurately measured by the major amblyoscope. Horizontal fusional reserves are more than vertical fusional reserve are (about 3–4 Δ). Convergence fusional reserves (about 25 Δ) are more than divergence fusional reserves (about 8 Δ for distance).

56 **A** = False **B** = False **C** = True **D** = True **E** = True

Binocular tests are to ensure that the visual acuities tested monocularly are similar, and that any residual refractive errors are equal in both eyes in a binocular condition. Binocular visual acuity is often better than that in each eye alone if the visual acuities in the eyes are nearly equal. Binocular tests may include the fogging technique in the binocular condition, dissociation of the eyes by various techniques (including the use of prisms, use of different coloured lenses) and comparing the results of the Duochrome test in binocular conditions. Tests may also include muscle imbalance.

57 **A** = False **B** = True **C** = True **D** = True **E** = True

The interpupillary distance varies according to the facial contour, and differs amongst individuals. It is about 2 mm shorter for females than males on average, and may be unequal in the two halves if there is significant facial asymmetry. The interpupillary distance should be measured both for distant and near vision.

58 **A** = True **B** = False **C** = True **D** = True **E** = True

Applying the general formula

$$F_d = \frac{F}{1 + dF}$$

where F_d = refractive power of the correcting lens placed d metres away from the eye (negative if moved towards the eye), F = refractive power of the correcting lens of the original lens in dioptres (+ve for convex lens, –ve for concave lens) and d = distance moved away from the eye (negative if moved towards eye).

If a contact lens is used

$$\begin{aligned} F &= 12\text{D} \\ d &= -15 \text{ mm } (-0.015 \text{ m}) \\ F_d &= 12/(1 + -0.015 \times 12) \\ &= 14.63\text{D} \end{aligned}$$

If a posterior chamber intra-ocular lens is used

$$\begin{aligned} F &= 12\text{D} \\ d &= -20 \text{ mm } (-0.02 \text{ m}) \\ F_d &= 12/(1+ -0.02 \times 12) \\ &= 15.79\text{D} \end{aligned}$$

59 **A** = False **B** = True **C** = True **D** = True **E** = True

Best form lenses for astigmatic lenses are considerably more complicated than spherical lenses, and often the ideal lens is unachievable. As bending of the lens is the main variable, equality of effective powers of the lens over its whole area cannot be achieved, and a compromise has to be made. Computers help in the design. Reflection from the bent lens is often a problem.

60 **A** = True **B** = True **C** = True **D** = True **E** = True

Accommodation changes the refractive state of the eye quickly and helps to focus objects at various distances. Hence, a cardinal symptom of abnormality of accommodation is blurring of vision. If the abnormalities are intermittent (e.g. accommodative fatigue), the symptoms would also be intermittent. Consistent excessive accommodative effort can lead to 'eye strain' and headaches. It usually affects either near vision (when there is deficient accommodation) or distant vision (when there is excessive accommodation) only.

TEST 3
Questions

1 Light

A has some of the properties of particles
B has some of the properties of waves
C needs a medium in which to propagate
D is an example of longitudinal waves
E travels in a direction perpendicular to the magnetic field

2 Dispersion

A is one of the properties of light
B is synonymous with refraction
C can be demonstrated by passing a white light through a triangular prism
D is the variation of index of refraction with frequency
E generally increases with the refractive index of the medium

3 Polarisation of light

A occurs if the light is monochromatic
B occurs if the electric fields of different light waves are in the same direction
C may occur when light passes through iodide crystals
D may occur by reflection
E may occur by scattering

4 A 1 cm high object is placed 10 cm in front of a concave mirror with a focal length of 8 cm.

A The image is inverted.
B The image is real.
C The image is 40 cm away from the mirror.
D The image is behind the mirror.
E The image is 2 cm in length.

5 In the emmetropic eye

A the total refractive power is about 59D
B the cornea and the lens contribute equally to the total refractive power
C the nodal point of the eye is at the anterior pole of the crystalline lens
D the distance between the anterior corneal surface and the retina is about 24.1 cm
E the first focal point is about 15.7 mm in front of the eye

6 Laser

A light is monochromatic
B light may have different polarisation from the original photon
C light is in phase with the original photon
D light is unidirectional
E light has an intensity of more than 1000 times that of sunlight in the spectral range of the laser beam

7 The following statements about astigmatism is/are correct

A It can be caused by pressure of the lids on the globe
B In regular astigmatism, the directions of greatest and least curvature are at 45° to each other
C Regular astigmatism can be corrected by a spherical lens
D In 'against the rule' astigmatism, the vertical meridian is more curved than the horizontal meridian
E Irregular astigmatism can be corrected by a combination of spherical and cylindrical lenses

8 In the conoid of Sturm produced by a lens system of +4DS/–2DC × 40°

A the circle of least diffusion is 33 cm from the lens
B addition of +1DS would move the proximal focal line 8.3 cm closer to the lens
C addition of +1DS would move the proximal focal line a longer distance than the distal focal line
D addition of –1DS would move the proximal focal line a longer distance than the distal focal line
E addition of –1DC × 40° will move the distal focal line away from the lens

9 The following statements about spherical lenses is/are true.

A The total refractive power equals the sum of the surface refractive power for thick lenses.
B Light parallel to the principal axis is converged onto the nodal point.
C Light passing through the nodal point is undeviated.
D The refractive power of a lens with focal length of 10 cm is 10D.
E The refractive power of a lens with focal length of –5 cm is 5D.

10 A convex lens of 3D is in contact with a concave lens of 2D homocentrically.

A The refractive power of the combination is independent of the distance between the two lenses.
B The refractive power of the combination is independent of the thickness of the lenses.
C The focal length of the combination is 5 m.
D The refractive power of the combination is 5D.
E The focal length of the combination is independent of whether the lenses are placed homocentrically.

11 A hypermetropic subject requires a spectacle correction of +12D. The back vertex distance is 15 mm.

A A contact lens of +9D results in under-correction.
B A contact lens of +13.5D results in over-correction.
C A contact lens of +16D results in over-correction.
D A spectacle lens of +13.5D results in over-correction if the back vertex distance is 10 mm.
E A spectacle lens of +11.25D results in under-correction if the back vertex distance is 10 mm.

12 A subject has a far point at 4 m and an amplitude of accommodation of 5D.

A The near point is about 19 cm in front of the eye.
B The near point cannot be determined.
C Addition of a –0.25D lens increases the range of accommodation.
D Addition of a +2D lens increases the amplitude of accommodation.
E Addition of a +1.25D lens moves the near point to 15.4 cm in front of the eye.

13 Myopia can be caused by

A excess shortening of the eye
B changes in the refractive index of the lens
C changes in the curvature of the cornea
D reading too closely for too long
E backward displacement of lens

14 The following statements about hypermetropia is/are true.

A In newborn babies, most hypermetropia are manifest.
B In old age, most hypermetropia are absolute.
C In newborn babies, most hypermetropia are latent.
D As age increases, manifest hypermetropia increases.
E As age increases, latent hypermetropia increases.

15 The following instruments are useful in the determination of the axis in astigmatism.

A The Maddox rod.
B The astigmatic fan.
C The stenopaeic slit.
D The cross-cylinder.
E The major amblyoscope.

16 When the eye is fully accommodated

A the ciliary muscle is in tension
B the ciliary body is closer to the lens
C the zonule is tensed
D the shape of the lens depends on its own elasticity
E the lens capsule is slack

17 The following statements about pathological axial myopia is/are true

A The anterior pole of the eye is most affected
B Posterior staphyloma may be a complication
C Myopic crescent is characteristically present
D Degeneration occurs in the central part of the fundus
E Degenerative changes in the vitreous are common

18 The following statements is/are true regarding refraction for subjects at different ages.

A Cycloplegics are generally required for children under the age of 10.
B Atropine should be used in all children under the age of 13.
C The near point for accommodation should be tested for a subject aged 35.
D Examination of the fundus is unnecessary for children under 5 years old.
E Muscle balance for near vision should be tested in those over the age of 50.

19 The following statements about presbyopia is/are true.

A The basic physiological process commences at about 40 years of age.
B The basic physiological process occurs earlier in hypermetropia.
C It increases throughout life.
D It occurs later in subjects with hypermetropia.
E It is manifested when the near point has receded beyond normal reading distances.

20 A biconcave lens has radii of curvature 8 cm on one surface A and 10 cm on another surface B. Assuming the refractive index of the lens is 1.5

A the refractive power at surface B is stronger than –6D
B the refractive power at surface B is weaker than –10D
C the refractive power at surface A is stronger than that at surface B
D the refractive power at surface A is –6.25D
E the refractive power of the lens is dependent on the refractive index of the lens

21 In the correction of ametropia

A a stronger concave lens is required for myopia if the lens is placed further away from the eye
B a stronger convex lens is required for hypermetropia if the lens is placed further away from the eye
C the retinal image is larger than in an emmetropic eye if a convex lens is placed 25 mm away from the eye to correct axial hypermetropia
D the retinal image is larger than in an emmetropic eye if a concave lens is placed 25 mm away from the eye to correct axial myopia
E the retinal image is larger than in an emmetropic eye if a concave lens is placed 7 mm from the eye to correct axial myopia

22 The following statements about aphakia is/are true

A Removal of the lens in an emmetropic eye requires correction by a spectacle convex lens of about 11D
B Removal of the lens in an emmetropic eye requires correction by a contact lens of less than 11D
C The retinal image of an aphakic eye is about 30% smaller than the emmetropic retinal image
D Unilateral aphakia is best corrected by spectacles
E It may be accompanied by astigmatism

23 The following statements about anisometropia is/are true.

A It occurs when both eyes are equally highly myopic.
B It occurs when one eye is highly myopic and the other emmetropic.
C Binocular single vision is impossible if anisometropia exceeds 1D.
D Correction by spectacle lenses often causes aniseikonia.
E An iseikonic lens is always effective.

24 Visual acuity is dependent on the following factors.

A Pupil size.
B Illumination of the target.
C Contrast with the background.
D Whether the subject is colour blind.
E Refractive error.

25 The following statements about the Duochrome test is/are true.

A It makes use of the spherical aberration of the eye.
B It makes use of the fact that green light is focused in front of red light on the retina.
C The under-corrected hypermetropic subject will see the green target more distinctly.
D The over-corrected myopic subject will see the red target more distinctly.
E It is an objective method of refraction.

26 In refraction, methods to refine objective tests include

A pinhole aperture
B Duochrome test
C fogging method
D stenopaeic slit
E binocular balance

27 Chromatic aberration

A is desirable in optical instruments
B is made use of in the Duochrome test
C may be reduced in lenses by combining different materials
D increases if material of high refractive index is used
E increases if material of high dispersive power is used

28 A +3D spherical lens placed in front of a subject causes neutralisation at two different meridians at 90° when the examiner is 66.6 cm away from the subject.

A The subject has astigmatism.
B The subject is 1.5D hypermetropic.
C The subject should be prescribed a 1.5D concave lens.
D The same neutralisation point will be reached at any other meridian.
E Subjective refraction need not be performed.

29 The following methods can be used to reduce accommodation in clinical refraction.

A Pilocarpine.
B Cyclopentolate.
C Gazing at a distant light source.
D Use of a positive spherical lens.
E Use of a cylindrical lens.

30 The following statements about the fogging method is/are correct.

A It is used to reduce accommodation.
B It is achieved by placing a convex lens in one eye only.
C It can be performed in less than one minute.
D It requires reduction of an added lens by 1D each time until maximum acuity is achieved.
E The minimum convex lens allowing full vision should be prescribed.

31 In retinoscopy, the following conclusion(s) can be drawn if there is a 'break' in the alignment between the reflex in the pupil and the band outside it.

A The power of the cylinder must be incorrect.
B Both the power and the axis of the cylinder must be incorrect.
C The true axis of astigmatism must be in the direction of the band in the pupil.
D The true power of astigmatism must be more positive than the trial cylindrical lens.
E The true spherical power must be more positve (convex) than the trial spherical lens.

32 Retinoscopy is carried out when large refractive errors are present.

A The examination should always commence assuming the subject is emmetropic.
B The reflex obtained on retinoscopy is usually faint when the refractive error is largely uncorrected.
C If the reflex is seen with no trial lens only when the examiner is very close to the subject, myopia is likely to be present.
D The examination may be performed with the refractive error indicated by the present spectacles as a starting point.
E Results obtained are always inaccurate.

33 The following is/are true regarding use of the astigmatic fan.

A It can be performed more effectively when the subject is corrected by a best vision sphere than if the subject is rendered 0.5D myopic.
B It can be performed more effectively when the subject is rendered 0.5D hypermetropic than if rendered 0.5D myopic.
C Negative cylinders should be added with the axis in the direction of the sharply-defined line.
D The absence of any sharply defined line is a sensitive test to exclude astigmatism.
E The presence of a sharply defined line is a sensitive test indicating astigmatism.

34 In indirect ophthalmoscopy

A the image is real
B the image is erect
C the image is magnified
D the amount of magnification/diminution is independent of the distance between the lens and the eye
E the image with a +20D lens is larger than with a +13D lens

35 The following statements is/are true regarding the use of indirect ophthalmoscope in diagnosing opacities in the refractive media.

A The opacities appear black.
B Opacities that continue to move when the subject's eye movement suddenly stops can be in the crystalline lens.
C Opacities that continue to move when the subject's eye movement suddenly stops can be in the vitreous.
D Opacities in the vitreous move in the opposite direction to the eye.
E Opacities in the anterior chamber move in the same direction to the eye.

36 The following statements about Risley's prism is/are true.

A It consists of two prisms of different power.
B The two prisms may be rotated in opposite directions by a knob.
C The maximum prismatic power is obtained when aligned base to base.
D It can produce different prismatic powers.
E The prismatic power is zero if the apex of one prism is directly in front of the base of the other.

37 The following statements is/are true regarding the measurement of prismatic power of a glass prism (refractive index 1.5).

A The angle of apparent deviation of 2° is about 1 Δ.
B A refracting angle of 4° is equivalent to about 2° angle of apparent deviation.
C An angle of apparent deviation of 2° is equivalent to about 4 Δ.
D An angle of apparent deviation of 3° is about 6 ∇.
E A refracting angle of 4° is equivalent to about 2 Δ.

38 During retinoscopy, the examiner leaned forward and found that the reflex moved in the same direction as the tilt of the mirror in two meridians. On leaning backward, the reflex moved in the same direction in the first meridian but in the opposite direction in a second meridian. The axis of cylindrical trial lens is in the direction of the second meridian.

A The refractive error is purely spherical.
B The subject has some degree of astigmatism.
C The neutralisation point was reached in both meridians.
D A positive cylindrical lens should be placed with the axis in the direction of the second meridian.
E The subject must be myopic.

39 Hruby lens

A is a biconcave lens
B forms a real image of the illuminated retina
C forms an erect image of the illuminated retina
D should be placed at least 1.5 cm away from the eye
E can be used without the slip-lamp

40 The following statements about the slit-lamp is/are true.

A The basic optics are similar to that of an astronomical telescope.
B The basic optics are similar to that of a high powered binocular microscope.
C A slit is placed between the two convex lenses.
D The foci of the illumination system and the viewing system coincide.
E It can be used to incorporate a three-mirror contact lens.

41 In Worth's four dot test

A a red lens is placed in front of the subject's left eye
B a plane untinted lens is placed in front of the subject's right eye
C the subject views a box containing two red lights
D the subject views a box containing two green lights
E the subject views a box containing one white light

42 The following about trifocal lens is/are correct.

A They are useful when the amplitude of accommodation is about 3D.
B They reduce some of the 'jump' effect of bifocals.
C The intermediate addition is usually half of the full addition for close work.
D They are useful for musicians.
E They are most suitable for subjects with anisometropia.

43 The following statements about trial lens used in refraction is/are correct.

A A convex lens can be in the form of a biconvex lens.
B A concave lens can be in the form of a biconcave lens.
C If a plano-convex lens is used, the plane surface of the lens should be close to the eye.
D If a plano-concave lens is used, the plane surface of the lens should be close to the eye.
E A trial lens should include a prismatic lens.

44 The following is/are true about the Galilean telescope.

A The eyepiece is convex.
B The objective lens is convex.
C The distance between the lenses can be determined by the focal lengths of the two lenses.
D It is less compact than the 'astronomy telescope' system.
E It produces an erect image.

45 The following is/are true regarding the use of a 20D convex lens as a magnifier.

A The magnification is 5 times if the object is 25 cm from the eye.
B The image is inverted.
C Moving the magnifier towards the eye would increase the field of vision.
D It utilises the principle of a loupe.
E A real image is formed.

46 The following statements is/are true regarding the fitting of contact lenses in subjects with astigmatism.

A A toric anterior curve lens is most suitable for those with pure lenticular astigmatism.
B Contact lenses are unsuitable for irregular astigmatism.
C A bitoric lens can be used for those with severe corneal astigmatism.
D A toric anterior curve lens is most suitable for those with severe corneal astigmatism.
E A subject with 0.5D astigmatism can be fitted with a posterior spherical curve lens.

47 Advantages of contact lenses over spectacles in monocular aphakia include

A increased visual field
B less spectacle diminution
C increased binocular single vision
D less ring scotoma
E less image distortion

48 The following statements regarding the statutory definition of blindness is/are true.

A The occupation of the persons concerned must be taken into account in deciding whether they are statutory blind.
B The mental and other physical conditions of the persons concerned must be taken into account in deciding whether they are statutory blind.
C The binocular corrected visual acuity measured by a Snellen Chart is the most important factor.
D A person with binocular visual acuity of less than 3/60 may be regarded as blind.
E A person with binocular visual acuity of 6/60 should never be regarded as blind.

49 The following statements is/are true regarding extended wear contact lenses.

A They should be changed every month.
B The number of new lenses used depends on the wearer's tear quality.
C A low Dk/L value is desirable.
D They should be cleaned every two weeks.
E The number of new lenses used depends on the wearer's environment.

50 The following are important factors determining contact lens fitting and wearing behaviour.

A Surface wettability.
B Gas permeability.
C Light transmission percentage.
D Elasticity.
E Water absorption.

51 'With the rule' astigmatism

A is correctable by a plus sphere at 90°
B is correctable by a minus cylinder with the axis at 90°
C is more common in children and young adults
D is frequently 'symmetrical' in both eyes so that the sum of the cylindrical axes in the two eyes is about 180°
E is a form of oblique astigmatism

52 Spherical aberration in spectacle lenses

A is less if a more concave surface is placed next to the eye
B increases with the power of the lens
C is reduced by aplanatic lenses
D renders aplanatic lenses essential for any refractive errors greater than 3D
E is reduced if the refractive power gradually increases towards the periphery from the centre

53 The following statements about relative convergence is/are correct.

A It is the difference between the range of convergence in the two eyes.
B It is the proportion of convergence exerted compared with the normal.
C It is the amount of convergence that can be made to vary when accommodation is kept constant.
D The negative portion should be kept higher than the positive portion.
E Prisms can be prescribed if the positive portion is persistently lower than the negative portion.

54 The following statements is/are true regarding relative prismatic effects when spectacle lenses are decentred.

A They are particularly large in anisometropia.
B They can cause discomfort for vertical imbalance of more than 1 Δ.
C If large, they may be associated with diploplia.
D They can be reduced by using a smaller lens.
E They are the differences of the prismatic effect between the two eyes when not in their primary positions.

55 The following statements is/are true regarding angle kappa.

A It is the angle between the anatomical and optical axes of the eye.
B Positive values cause the appearance of exotropia (divergent squint).
C Negative values cause the appearance of esotropia (convergent squint).
D Abnormal values are best corrected surgically.
E A large negative value can be caused by high myopia.

56 Regarding binocular vision

A the failure of simultaneous perception implies there must be suppression
B the lack of stereopsis implies the total lack of binocular vision
C the presence of stereopsis implies the presence of simultaneous perception
D the presence of fusion implies the presence of simultaneous perception
E the presence of fusion implies the presence of stereopsis

57 Tinted glasses

A are never indicated
B may be useful in tropical countries
C should never be used in myopia
D are useful in albinism
E are sometimes useful in aniridia

58 *Pseudomonas* infection in contact lens wearers

A is usually self-limiting
B is more common in hard-lens wearers
C may be exacerbated by the use of topical steroid eye drops
D may present with severe ocular pain
E requires a corneal scrape to be taken before treatment

59 The following about the AC/A ratio is/are correct.

A It is the ratio of the amount of convergence (in Δ) for every D change in accommodation.
B A low value may cause divergent squint when viewing near objects.
C It can be reduced by correction of refractive error.
D It can be increased by orthoptic exercises.
E A high value may cause convergent squint when viewing near objects.

60 Compared with spectacle prescription, treatment of anisometropia by contact lenses

A reduces the prismatic effect in axial anisometropia
B reduces the prismatic effect in refractive anisometropia
C reduces the aniseikonia in axial anisometropia
D reduces the aniseikonia in refractive anisometropia
E increases the monocular visual acuities

TEST 3
Answers

1 **A** = True **B** = True **C** = False **D** = False **E** = True

Light can be described both by a particle model (the photon) and by a wave model (the electromagnetic wave). Waves can be divided into longitudinal waves, where the molecules of the medium are displaced backwards and forwards parallel to the direction of travel (e.g. sound waves) and transverse waves, where the disturbance created by the wave is perpendicular to the direction of travel of the wave. A longitudinal wave needs a medium through which to travel. Light is an electromagnetic wave where the disturbance is a time-varying electric and magnetic field. The directions of the electric field, magnetic field and the direction of travel are all perpendicular to one another. A transverse wave does not require a medium to propagate.

2 **A** = True **B** = False **C** = True **D** = True **E** = True

Dispersion is the variation of the refractive index with the frequency of light. The refractive index is usually higher the higher the frequency (or the lower the wavelength). Hence, the refractive index is higher for blue light than red light. This can be demonstrated by the production of a spectrum by directing a white light into a triangular prism. Generally speaking, the higher the refractive index of a medium, the higher its dispersion. Hence, a powerful lens made from materials of high refractive index is more likely to produce chromatic aberration.

3 **A** = False **B** = True **C** = True **D** = True **E** = True

Light is an electromagnetic wave with the electrical and magnetic components at right angles to the direction of travel. If more than one light wave travels with the electrical components in different directions, then the light beam is unpolarised. If the electrical components are orientated in the same direction, the light beam is polarised. Unpolarised light may be polarised by transmission, reflection or scattering.

In transmission through iodide crystals, light with the electric field parallel to the direction of the iodide molecules would be absorbed, and only that at rights angle to the iodide molecules can pass through.

Light travelling from a medium to a denser medium would be partly refracted and partly reflected. If it is so arranged that the reflected light is perpendicular to the refracted light, the reflected light would be polarised, as it is impossible for the electrical component of the light to be in the same direction as the direction of propagation. This is the basic principle of how sunglasses reduce glare (see Fig. 3.1).

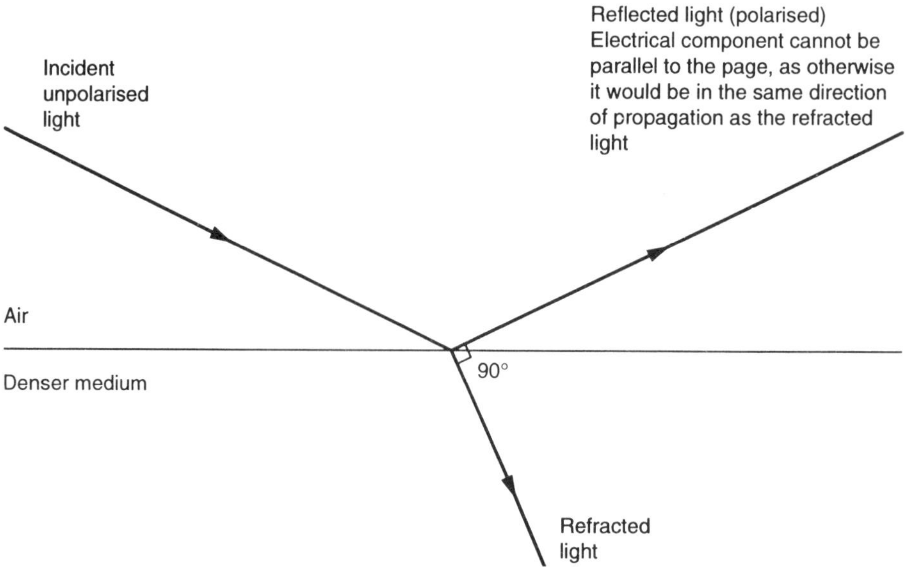

Figure 3.1 Polarisation by reflection.

Light can also be polarised by scattering by the atmosphere (reflection by small particles) in much the same principle as polarisation by reflection.

4 **A** = True **B** = True **C** = True **D** = False **E** = False

The problem can be tackled either by a ray diagram or by formula (see Fig. 3.2). A ray of light parallel to the axis of the mirror is reflected and passes through the focal point. Also, a ray of light passing through the centre of curvature (C) is reflected along itself. Hence, the image is enlarged and inverted. The size can be measured, or it can be calculated from the formula

$$\frac{1}{u}+\frac{1}{v}=\frac{1}{f}$$

Hence

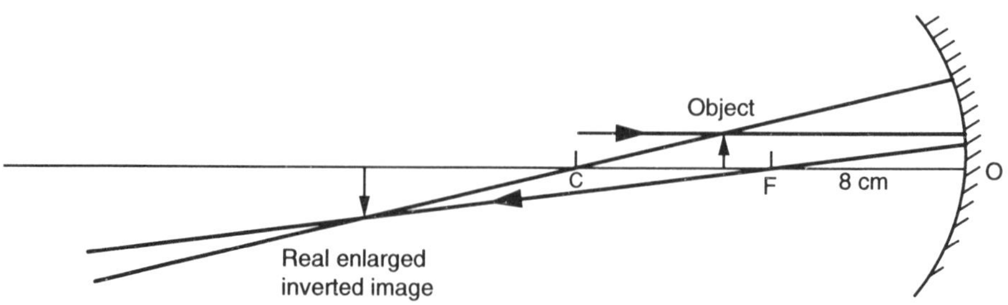

Figure 3.2 Reflection by concave mirror. F= focal point, C = centre of curvature of mirror.

$$\frac{1}{0.1}+\frac{1}{v}=\frac{1}{0.08}$$

$$v = (1/2.5)\ \text{m} = 40\ \text{cm}$$

The magnification is v/u = 4 times.

Similar situations are frequently met in ophthalmic instruments employing concave mirrors.

5 **A** = True **B** = False **C** = False **D** = True **E** = True

The total refractive power of the eye is about 58.6D, of which more than two thirds is contributed by the cornea, the rest being contributed by the lens. This is because there is a greater difference between the refractive index between air and the cornea (1 and 1.38 respectively) than the difference between the lens and the aqueous/vitreous humour (1.41 and 1.3 respectively).

In the reduced eye, the nodal point lies on the posterior capsule of the lens. Hence, the posterior capsular cataract results in a greater reduction of visual acuity than other cataracts. The first focal point is about 15.7 mm in front of the cornea. This is important as a spectacle lens placed at this point results in a corrected image size the same as the emmetropic image size. The second focal point is 24.1 mm behind the cornea (i.e. at the retina).

6 **A** = True **B** = False **C** = True **D** = True **E** = True

A laser beam is unidirectional to within 1 minute of arc. The intensity is extremely high and within the spectral range of the laser, can be several thousand times that of sunlight. It is almost monochromatic. Laser light is coherent, being of the same polarisation and exactly in phase with the original photon.

7 **A** = True **B** = False **C** = False **D** = False **E** = False

Astigmatism is often caused by the curvature of the cornea being abnormal, so that it is no longer spherical. **Regular astigmatism** occurs when the cornea has its directions of greater and least curvature at 90° to each other. The most common form is when the vertical meridian is more curved, probably due to the pressure of the lids on the globe 'squeezing' it. This is called **'according to the rule' astigmatism**. Conversely, when the horizontal meridian is more curved, it is called **'against the rule' astigmatism**. When the meridians of astigmatism are not at right angle to each other, it is called **irregular astigmatism**, and cannot be totally corrected by a spectacle lens. A contact lens may partially correct the defect.

8 **A** = True **B** = False **C** = False **D** = False **E** = True

The power at the focal lines are 4D and 2D. At the circle of least diffusion, the power is (4 + 2)/2 = 3D. Hence, the circle of least diffusion is 1/3 m = 33 cm away from the lens.

Addition of +1DS increases the power at the proximal focal line to (4 + 1) = 5D. Hence, it is 1/5 = 20 cm away from the lens. The proximal focal line has moved (25 – 20) cm = 5 cm closer to the lens.

Similarly, addition of +1DS increases the power at the distal focal line to 3D. It is 1/3 m = 33.3 cm away from the lens. Hence, it moves (50 – 33.3) cm = 16.7 cm closer to the lens. The proximal focal line moves a shorter distance than the distal focal line.

Addition of –1DS would produce a lens system of +3DS/–2DC × 40°. The powers at the focal lines are 3D and 1D. The proximal and distal focal lines are 33.3 cm and 100 cm away from the lens respectively. Hence, the proximal focal line is moved (33.3 – 25) cm = 8.3 cm, and the distal focal point (100 – 50) = 50 cm away from the lens.

Addition of –1DC × 40° would produce a lens system of +4DS/–3DC × 40°. The distal focal line would have a power of 1D, and is 1 m away from the lens. Hence, it has moved 100 – 50 = 50 cm away from the lens.

9 **A** = False **B** = False **C** = True **D** = True **E** = False

The optics of thin lenses are considerably simpler than that of thick lenses. In ophthalmology, most of the lenses used are thin lenses. In thin lenses, the total refractive power equals the sum of the surface refractive power. The line joining the radius of curvature of the surfaces is called the principal axis. The principal axis meets the middle of the lens at the nodal point (N). There are two centres of curvature (R1 and R2) and two principal foci (F1 and F2) on either side of the lens. The distance between the principal focus and the nodal point is the focal length (*f*). In thin lenses, the radius of curvature is twice the focal length.

Light passing through the nodal point is undeviated (*X*). Light parallel to the principal axis converges onto the second principal focus (*Y*) By the principle of the reversibility of light, light passing through the first principal focus emerges parallel to the principal axis (see Fig. 3.3).

The refractive power of a lens is measured in dioptres.

Power (in dioptres) = 1/second focal length (in metres).

The stronger lens has a shorter focal length.

The refractive power of a lens with focal length of 10 cm (0.1 m) = 1/0.1 = 10 D. The refractive power of a lens with focal length of –5 cm (0.05 m) = 1/–0.05 = –20D.

10 **A** = False **B** = False **C** = False **D** = False **E** = False

A system of lenses is said to be homocentric when all the component lenses are centred on a common optical axis. If several lenses are placed homocentrically and close to each other, the total refractive power of the lenses is equal to the sum of the individual lenses, taking the convex lens as (+) and the concave lens as (–).

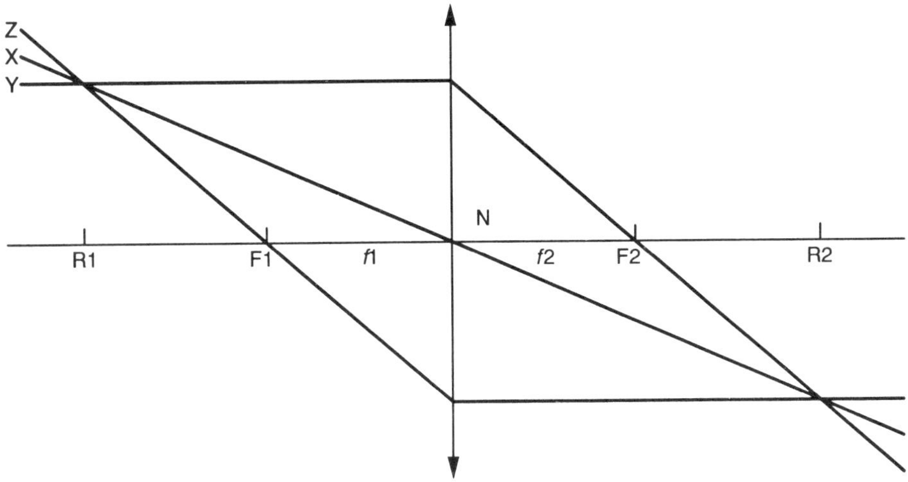

Figure 3.3 The optics of thin lenses.

If a convex lens of +3D is placed homocentrically with a concave lens of –2D, the total refractive power of the combination is 1D. Hence, the focal length of the combination is 1/1 = 1 m.

The above calculations are true only if the lenses are thin, and are close to each other.

11 **A** = True **B** = False **C** = True **D** = True **E** = True

From the general equation

$$F_d = \frac{F}{1 + dF}$$

where F_d = refractive power of the correcting lens placed d metres away from the eye (negative if moved towards the eye), F = refractive power of the correcting lens of the original lens in dioptres (+ve for convex lens, –ve for concave lens) and d = distance moved away from eye (negative if moved towards eye).

If a contact lens is used

$$\begin{aligned} F &= 12\text{D} \\ d &= -0.015 \text{ m} \\ F_d &= 12/(1 + (-0.015 \times 12)) \\ &= 14.63\text{D} \end{aligned}$$

If the back vertex distance is 10 mm

$$\begin{aligned} F &= 12\text{D} \\ d &= -0.005 \text{ m} \\ F_d &= 12/(1 + (-0.005 \times 12)) \\ &= 12.76\text{D} \end{aligned}$$

12 **A** = True **B** = False **C** = True **D** = False **E** = True

A subject with a far point of 4 m would be (1/4) = 0.25D myopic. Hence, the refractive power for the near point is (5 + 0.25) = 5.25D. The near point is (1/5.25) m = 19 cm. Addition of a –0.25D lens would correct the myopia, and the far point is infinity, hence the range of accommodation is increased. The amplitude of accommodation remains unchanged whatever lens is in front of the eye. Addition of a +1.25D lens would render the subject (1.25 + 0.25) = 1.5D myopic. Hence, at the near point, the power of refraction is (1.5 + 5) = 6.5D. The near point is (1/6.5) m = 15.4 cm.

13 **A** = False **B** = True **C** = True **D** = False **E** = False

Myopia can be caused by excessive lengthening of the eyeball. There is no evidence that the length of time spent in reading is associated with development of myopia. The forward dislocation of lens is associated with myopia, as the effective power of the convex lens increases when it moves forward from the nodal point. Conversely, a concave lens is more effective the more it moves towards the nodal point of the eye.

14 **A** = False **B** = True **C** = True **D** = True **E** = False

In newborn babies, the ciliary muscle tone is high. Most of the hypermetropia is latent, and is only revealed by the use of cycloplegics. In old age, the lens nucleus becomes harder, and most hypermetropia becomes manifest. Also, with increasing of age, the amplitude of accommodation decreases, and most of the hypermetropia becomes absolute.

15 **A** = False **B** = True **C** = True **D** = True **E** = False

The axis of astigmatism can be determined objectively by retinoscopy, or subjectively by the stenopaeic slit, the astigmatic block or fan, and the cross-cylinder. The stenopaeic slit contains a slit mounted on a trial lens so that light in one meridian only can enter the eye. When the spherical component of the refractive error is corrected, the vision would be best if the slit aperture is in line with the astigmatic axis of the eye. Similarly, in the astigmatic fan, the line which appears to be clearest (or darkest) would be in line with the astigmatic axis of the eye. See Test 1 Question 25 for the cross-cylinder. The major amblyoscope is mainly used to measure squints and binocular single vision.

16 **A** = True **B** = True **C** = False **D** = True **E** = False

See Test 1 Question 18.

17 **A** = False **B** = True **C** = True **D** = True **E** = True

See also Test 2 Question 17. In pathological axial myopia, the posterior pole of the eye is much more affected by the increase in the length of the eye than the anterior pole. This is probably due to degeneration of the sclera. In severe cases, the sclera may bulge out at the posterior pole to form a posterior staphyloma. Degeneration occurs both in the central and the peripheral parts of the retina.

Degeneration in the periphery of the retina leads to retinal tears and detachment. Degeneration changes in the vitreous are common, causing 'floaters'.

18 **A** = True **B** = False **C** = True **D** = False **E** = True

Children generally require cycloplegics for refraction, as they have a large amplitude of accommodation and are unable to relax voluntarily. Atropine should probably be used in the very young (say, under 5), but in older children, cyclopentolate should be sufficient. The fundus should always be examined in all subjects. In children under age of 5, retinoblastoma may be a cause for refractive error.

The near point for accommodation and convergence should be tested in subjects over the age of, say, 30. The muscle balance for near vision should generally be performed above the age of 40.

19 **A** = False **B** = False **C** = True **D** = False **E** = True

Presbyopia is caused by the gradual reduction in the amplitude of accommodation as age increases, and commences from birth throughout life. In children, the amplitude of accommodation is more than 14D, and it gradually decreases to 10D aged 20, 7D aged 30, 4.5D aged 40, 2D aged 50, and less than 1D aged 60. It becomes manifest when the near point recedes beyond a comfortable reading distance. It becomes manifest earlier in hypermetropic subjects as part of their accommodation is used to overcome their hypermetropia.

20 **A** = False **B** = True **C** = True **D** = True **E** = True

Using the formula

$$\text{Power of surface (in dioptres)} = (n_1 - n_2)/r$$

where n_1 = refractive index of lens, n_2 = refractive index of air and r = radius of curvature of lens in m (+ve if convex, –ve if concave)

$$\text{Refractive power at surface B} = (1.5 - 1)/-0.1 = -5\text{D}$$
$$\text{Refractive power at surface A} = (1.5 - 1)/-0.008 = -6.25\text{D}$$
$$\text{Total power of lens} = (-5) + (-6.25) = -11.25\text{D}$$

21 **A** = True **B** = False **C** = True **D** = False **E** = True

When the spectacle lens is moved away from the eye, the image also moves forwards. In uncorrected myopia, the image is formed in front of the retina, and hence if the lens is moved forwards, a stronger concave lens is required to move the image back onto the retina. Conversely, in uncorrected hypermetropia, the image is behind the retina, and the convex lens for correction attempts to bring it forwards onto the retina. Hence, if the correcting convex lens is moved forwards, the image also moves forwards, and a weaker convex lens is needed.

Relative spectacle magnification is the ratio of corrected ametropic image size to emmetropic image size. In axial myopia or hypermetropia, the corrected ametropic image size is equal to the emmetropic image size. (The relative spectacle magnification is 1.) In axial myopia, if the spectacles are worn nearer to the eye than the anterior focal point (about 15 mm from the cornea), the image size is increased. Conversely, if they are worn more than 15 mm from the cornea, the image size is reduced. The opposite is true for hypermetropia.

22 **A** = True **B** = False **C** = False **D** = False **E** = True

Correction of aphakia after surgical removal of a lens in an emmetropic eye by spectacles requires a convex lens of about 11D, although it would require a higher correction by a contact lens or an intra-ocular lens. In hypermetropia, the image is behind the retina. Hence, the further forward is the correction lens, the further forwards would be rays of light brought onto the retina. The retinal image of an aphakic eye is about 30% larger than an emmetropic eye. Astigmatism is sometimes present due to the surgical section on the cornea or the corneo-scleral junction. Unilateral aphakia is best corrected by an intra-ocular lens if binocular single vision is desired.

23 **A** = False **B** = True **C** = False **D** = True **E** = False

Anisometropia occurs when the refractive errors of the two eyes show a significant difference. Some binocular single vision is possible for low grades of anisometropia, but eye-strain is a common complaint, and binocular single vision becomes impossible in high grades of anisometropia. Correction with spectacles usually causes differences in retinal image size (aniseikonia), and correction with iseikonic lenses are not always effective. Contact lenses may be more effective.

24 **A** = True **B** = True **C** = True **D** = False **E** = True

The visual acuity depends on pupil size. An increase in pupil size would increase spherical aberration while decreasing diffraction. Hence, there is an optimal pupil size (of about 2.5 mm). Visual acuity increases with increasing contrast with background and illumination of the target. Colour blindness generally does not affect visual acuity. The presence of refractive error would decrease visual acuity.

25 **A** = False **B** = True **C** = True **D** = False **E** = False

The Duochrome test makes use of the chromatic aberration of the eye. Blue and green lights are refracted to a greater degree than red light, and hence are focused in front of the red light on the retina. In the Duochrome test, Snellen letters illuminated by red and green light in two separate rows are shown to the subject. If the subject is emmetropic, both rows of letters appear equally clear. If the subject is myopic (i.e. under-corrected myopia or over-corrected hypermetropia), the letters illuminated by red light would appear clearer. Conversely, if the subject is hypermetropic (i.e. over-corrected myopia or under-corrected hypermetropia) the letters illuminated by green light would appear clearer.

26 **A** = False **B** = True **C** = True **D** = True **E** = True

A pinhole aperture is useful in estimating the final visual acuity. However, it is not used to refine objective tests. The stenopaeic slit can be used to refine the astigmatic axis by putting it with the best vision sphere, and gradually rotating it until the best visual result is obtained. The Duochrome test is used to refine the spherical correction, and binocular tests are usually finally performed to ensure that the visual results obtained uniocularly are obtained binocularly, and that the visual acuities in both eyes are balanced.

27 **A** = False **B** = True **C** = True **D** = False **E** = True

Chromatic aberration is caused by the fact that refractive indices vary in different wavelengths of light (higher in blue light and lower in red light). Hence, a series of images of different colours is formed when white light passes through a spherical lens, similar to the production of a spectrum with a prism. Chromatic aberration increases if a material of high dispersive power is used (i.e. when the difference in refractive index for different lights are greater), but it is independent of the actual refractive index of the material.

Chromatic aberration is undesirable in optical instruments as it interferes with the production of clear images. In an achromatic lens system, lenses of varying material (with different dispersive and refractive power) are combined to neutralise the dispersion while preserving the total refractive power. The Duochrome test exploits the chromatic aberration of the eye to test for slight myopia or hypermetropia by asking the subject to compare the visual effect of letters mounted on red and green illuminated lights.

28 **A** = False **B** = True **C** = False **D** = True **E** = False

The refractive error of the subject can be determined by subtracting working distance from the lens used (lens used – working distance refractive power).

The working distance of 66.6 cm is equivalent to (1/0.66) = 1.5D.
The subject is (+3) – 1.5 = +1.5D. (1.5D hypermetropic).

The neutralisation at the two meridians are the same. Hence, there is no astigmatism.

Subjective refraction should always be performed to confirm the findings of the retinoscopy results.

29 **A** = False **B** = True **C** = True **D** = True **E** = False

During subjective refraction, it is important to minimise accommodation for the refractive error to be determined accurately. For infants and young children, it is best to use cycloplegic agents, such as cyclopentolate. In adults, the fogging technique can be used with a positive spherical lens. When the subject is rendered myopic by a positive convex lens, the image is moved in front of the retina. Further accommodation would move the image even further forwards. Hence, it would have the effect of discouraging accommodation by the subject.

The positive spherical lenses are gradually removed until the best visual acuity is achieved. For cooperative adults, the instruction to gaze at a distant light source with the other eye may be adequate.

30 **A** = True **B** = False **C** = False **D** = False **E** = False

The fogging method is used to encourage relaxation of accommodation without using cycloplegics. After objective refraction, both eyes are blurred by the addition of a +2D convex lens to the correcting lenses, so that the visual acuity is less than 6/60. The subject should look at distant objects to relax the accommodation. The convex lens in one eye should be reduced slowly by 0.5D each time, until the best visual acuity is obtained. The maximum convex lens that would allow the best visual acuity is the correction in the absence of accommodation, and should be prescribed.

31 **A** = False **B** = False **C** = False **D** = False **E** = False

A 'break' in the alignment between the reflex in the pupil and the band outside it implies that the axis of the cylinder is incorrect, and the true axis is generally between that in the pupil and the band outside it. The powers of the cylindrical lens and the spherical lens do not affect the break phenomenon, except that the spherical correction must be approximately correct before the 'break' phenomenon occurs.

32 **A** = False **B** = True **C** = True **D** = True **E** = False

When there is large refractive error, one should estimate the approximate refractive error either with the present spectacles, or by trial of some high-powered lens (e.g. +5D, +10D, +15D, −5D, −10D, −15D). The retinoscopy reflex is usually faint when the refractive error is large, but usually becomes brighter when the refractive error is corrected. However, accurate results can be obtained.

If the reflex is seen with no trial lens only when the examiner is close to the subject, myopia is likely to be present. The working distance is very small, and a large positive spherical component needs to be subtracted from the power of the trial lens.

33 **A** = False **B** = False **C** = False **D** = False **E** = True

An astigmatic fan should be used when the subject is rendered 0.5D myopic by placing a 0.5D convex lens in front of the eye. The presence of hypermetropia encourages accommodation. In the absence of astigmatism, the subject would not see any line more sharply defined than others. However, this is not a sensitive test, as some subjects lack this power of observation. The presence of a sharply defined line (or a group of sharply defined lines), on the other hand, is a sensitive test indicating astigmatism. To find the power of the cylinder, negative cylinders should be added with the axis at right angles to the sharply defined lines until all the lines in the astigmatic figure appear uniform.

34 **A** = True **B** = False **C** = True **D** = False **E** = False

See Fig. 3.4.

In indirect ophthalmoscopy, a powerful convex lens (+20D or +13D) is held in front of the patient's eye, and a powerful illuminating light source is provided by a head lamp worn by the observer. A real, inverted, laterally inverted image is formed between the lens and the observer.

The magnification is about 3 times when a 20D lens is used, and 5 times when a +13D lens is used.

35 **A** = True **B** = False **C** = True **D** = True **E** = True

Any opacities in the refractive media would stop the light from being reflected from the retina, and hence they appear black. Opacities that continue to move when the subject's eye movement suddenly stops must be in a fluid, and can be in the aqueous or in the vitreous only. Objects at the plane of the pupil would appear stationary when the eye is moved. Objects behind the pupillary plane would appear to the examiner to move in the opposite direction, while objects in front of the pupillary plane would appear to move in the same direction as eye movement.

36 **A** = False **B** = True **C** = True **D** = True **E** = True

A Risley's prism consists of two prisms of the same power. It is arranged so that by turning a knob, both prisms are rotated symmetrically in opposite directions. The total prismatic power depends on the alignment of the two prisms. If the prisms are arranged with the base of one in front of the base of another, the resultant prismatic power is additive. Conversely, if the base of one is in front of the apex of the other, the resultant prismatic power is zero. Varying prismatic power can be obtained by placing the prisms in different alignment.

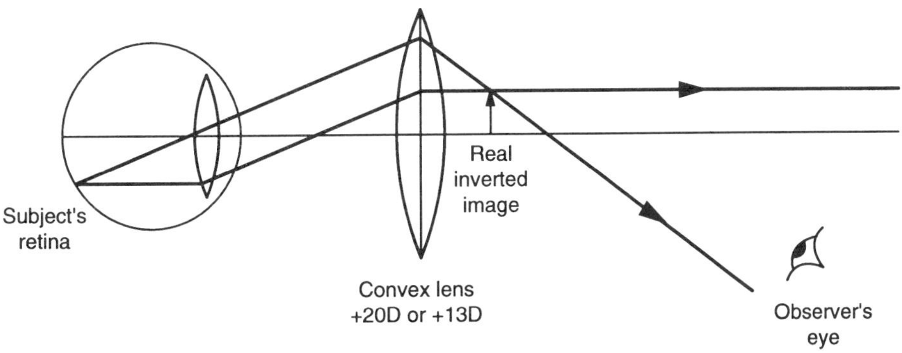

Figure 3.4 The indirect ophthalmoscope.

37 **A** = False **B** = True **C** = True **D** = True **E** = False

The angle of deviation in an ophthalmic glass prism is about half the refracting angle. Also, a prism of refracting angle 1° has 1 Δ power, and produces an apparent angle of deviation of (1/2)°. 1 ∇ is approximately the same (very slightly larger) than 1 Δ.

38 **A** = False **B** = True **C** = False **D** = True **E** = False

In one meridian, the reflex moves in one direction on leaning forwards and in the opposite direction on leaning backwards. Leaning forwards and backwards changes the working distance slightly. Hence, in this meridian, the point of neutralisation was achieved. In the other meridian, the direction of the reflex moves in the same direction both on leaning forwards and backwards. Hence, the neutralisation point has not yet been achieved.

As the neutralisation points are different in the two meridians, there must be astigmatism. As no information was given about the lens inserted in front of the subject's eye, no conclusion can be drawn about the refractive error.

The neutralisation point for the second meridian was reached at some working distance between the two distances tested. The neutralisation point for the first meridian was not between two working distances tested. The patient must be rendered more myopic with a positive cylindrical lens to obtain an against movement when the observer leans backwards.

39 **A** = False **B** = False **C** = True **D** = False **E** = False

Parallel light entering the eye is focused on the posterior part of the lens. Hence, the slit lamp alone cannot be used to view the structure in the posterior segment of the eye. However, the light can be made more diverging by using a concave lens in front of the eye.

Hruby lens is a plano-concave lens attached to the chin rest of the slit lamp via a holder. It is about –58.6D. It forms a virtual, erect, diminished image of the retina, so that it lies within the focal range of the slit lamp microscope. In order to maximise the field of view, it should be placed as close to the cornea as possible, with the concave surface facing the patient's eye.

40 **A** = False **B** = True **C** = False **D** = True **E** = True

The basic optics of the slit lamp are similar to the binocular compound microscope, which consists of two compound microscopes mounted at a slight angle (about 14°) to each other, giving a stereoscopic view. One compound microscope consists of two convex lenses, the objective and eyepiece. There is also an illumination system with the vertical slit of light focused on the point where the microscope is focused. The distance between the objective lens and the subject's eye is longer than the ordinary compound microscope to allow for such procedures as applanation tenometry, placing a Hruby lens or three-mirror contact lens, and removing foreign bodies.

41 **A** = False **B** = False **C** = False **D** = True **E** = True

Worth's four-dot test is used to test for abnormality of fusion and abnormal retinal correspondence. A **R**ed lens is placed in front of the **R**ight eye and a green lens is placed in front of the left eye. The subject views a box with one red, two green and one white light. Since only red light can pass through red lens, and only green light can pass through green lens, the red light must be seen by the right eye, and the two green lights by the left eye. If both eyes are used to see the white light, the light will be seen as white. However, if the right eye is suppressed, the white light will be seen as green by the left eye.

42 **A** = False **B** = True **C** = True **D** = True **E** = False

Trifocal lenses consist of a main portion for distance vision, and two added portions for intermediate and near vision. They are only useful if the amplitude of accommodation has decreased to such an extent that intermediate distance becomes blurred. They are not useful for someone with 3D of accommodation, as s/he can see objects at 50 cm (keeping one third of accommodation 1D as reserve). They reduce some of the prismatic jump effect of bifocals. They are useful for the musician who needs to read music at an intermediate distance. They are not suitable for subjects with anisometropia, as the vertical prismatic effect is not well tolerated.

43 **A** = False **B** = False **C** = True **D** = False **E** = True

A trial lens used in refraction should include positive and negative spherical lenses, positive and negative cylindrical lenses, prisms, plano lenses, opaque discs, pinholes, stenopaeic slit, and Maddox rods. If possible, red and green glass can be included.

Spherical lenses must be in the form of plano-convex or plano-concave lenses, as they are used in spectacles. When a plano-convex lens is used, the plane surface should be close to the eye. When a plano-concave lens is used, the concave surface should be close to the eye.

44 **A** = False **B** = True **C** = True **D** = False **E** = True

Also see Test 4 Question 45. The Galilean telescope is similar to the astronomical telescope, except that the eyepiece is concave, and the distance between the two lenses equals the difference between the focal lengths of the two lenses. Hence, it is more compact than the astronomical telescope system. It produces an erect, magnified image.

It can be used as a magnifying visual aid, usually mounted in a spectacle frame. A magnification of 3 times for distance and 8 times for near vision can be achieved.

45 **A** = True **B** = False **C** = False **D** = True **E** = False

A loupe is a convex lens of focal length shorter than 25 cm placed so that the object is placed at the anterior focal point of the lens. The angle subtended at the retina would be larger, and the angular magnification is (power of lens/4) compared with the unaided eye with the object at 25 cm from the eye.

Hence, the magnification is 20/4 = 5 times if a 20D lens is used. Moving the magnifier towards the eye would move the object away from the anterior focal point of the lens. The image is virtual and erect.

46 **A** = True **B** = False **C** = True **D** = False **E** = True

Contact lenses are used for both regular and irregular astigmatism, while spectacles lenses are suitable only for regular astigmatism.

Corneal astigmatism of less than 1.5D may be corrected by a posterior spherical curve lens alone. In severe corneal astigmatism, a posterior toric curve lens is needed to correct the astigmatism. However, there will be induced astigmatism from the lens–tear surface. Hence, a bitoric lens (with both toric anterior and posterior surfaces) may be needed.

In pure lenticular astigmatism, the cornea is spherical. Hence, a spherical posterior curve lens with a toric anterior surface to correct the astigmatism should be used.

47 **A** = True **B** = False **C** = True **D** = True **E** = True

The advantages of contact lenses over spectacles in monocular aphakia includes increase in visual field, less spectacle magnification, increased binocular single vision, less spherical aberration with ring scotoma and image distortion, less prismatic effect, and of lighter weight.

48 **A** = False **B** = False **C** = True **D** = True **E** = False

The statutory definition of blindness for the purposes of registration is that the person is 'so blind as to be unable to perform *any* work for which eyesight is essential'. Their occupation and their mental or other physical conditions must not be taken into account. The binocular corrected visual acuity is the most important factor. In general, a person with a binocular corrected visual acuity of less than 3/60 should be regarded as blind. A person with visual acuity of more than 6/60 may be regarded as blind only if the field of vision is markedly contracted, especially in the lower part of the visual field. A person with binocular visual acuity of between 3/60 and 6/60 may be regarded as blind if their visual acuity is considerably contracted.

49 **A** = False **B** = True **C** = False **D** = False **E** = True

Extended wear contact lenses should have a high Dk/L ratio, and should be of good tensile strength to last for about 2 weeks' wear. The lenses are disposable after use, and should not be cleaned. They should be stored in solutions as provided by the manufacturer. The number of new lenses used depends on the subject's tear flow and their environment. However, the lenses should generally be changed every one to two weeks.

50 **A** = True **B** = True **C** = True **D** = True **E** = True

Factors important in the fitting and wearing behaviour of a lens include surface wettability, stress and strain, gas permeability, refractive index, water absorption, light transmission percentage, and material purity.

51 **A** = False **B** = False **C** = True **D** = True **E** = False

'With the rule' astigmatism may be caused by the lid squeezing the globe. It may therefore increase in curvature vertically. It can be corrected by a plus **cylinder** at 90° or a minus cylinder at 0°. It appears to be more common in children and young adults, and tends to become 'against the rule' as age increases. The astigmatisms are frequently 'symmetrical' in that they are mirror images of each other and the sum of their cylindrical axes approximates 180°. Oblique astigmatism usually refers to astigmatism with the cylindrical axis more than 25° to the horizontal or the vertical.

52 **A** = True **B** = True **C** = True **D** = False **E** = False

Spherical aberration is less if a more concave surface of the lens is placed next to the eye and a more convex surface is placed away from the eye. To eliminate spherical aberration further, aplanatic lenses can be used. These are lenses whose refractive power gradually reduces from the centre to the periphery to form an aspherical surface. Since the eye has a mechanism to reduce spherical aberration and it is expensive to make aplanatic lenses, this form of lens is not generally required unless very strong lenses are used, as in aphakia.

53 **A** = False **B** = False **C** = True **D** = False **E** = True

Relative convergence is the amount of convergence which can vary while accommodation is kept constant. For comfort, the positive portion should exceed the negative portion, and the middle range should generally be used. Prisms can be prescribed to help with convergence deficits or excess.

54 **A** = True **B** = True **C** = True **D** = True **E** = True

The relative prismatic effect is the difference between the two eyes of the spectacle prismatic effect when the eyes are not in the primary positions. Since the prismatic effect = power of lens × decentration, it is particularly large in anisometropia (where the power of the lens differs considerably between the two eyes). Using a smaller lens may reduce the amount of decentration. The decentration gives rise to blurred vision, and a relative prismatic effect gives rise to diplopia. A vertical prismatic effect of more than 1 Δ would not be tolerated.

55 **A** = True **B** = True **C** = True **D** = False **E** = True

Angle kappa is the angle between the optical axis and the anatomical axis of the eye. Corneal light reflex is usually nasal to the centre of the cornea, as the macula is slightly temporal to the optic disc. Hence, a positive angle kappa gives the appearance of exotropia. Conversely, a negative angle kappa results in light reflex displaced temporally, and gives the appearance of esotropia. This usually results from high myopia, or rarely, retinal pathology. No action need be taken except to correct the underlying cause.

56 **A** = False **B** = False **C** = True **D** = True **E** = False

There are three levels of binocular vision. The first level is simultaneous perception, the ability to see dissimilar objects presented to each eye simultaneously. An example is to present the picture of a flower to one eye and a vase to another. Lack of simultaneous perception implies either suppression or amblyopia. The second level of binocular vision is fusion, the ability to combine two similar images that differ in some small details. The presence of fusion implies the presence of simultaneous perception. The third level of binocular vision is stereopsis, the ability to perceive, from two pictures taken at a slightly different angles, a three-dimensional image. Stereopsis can be tested either with the Titmus test (where the eyes are dissociated by polaroid glasses, and three-dimensional images of a fly, circles, or animals can be construed from polaroid pictures) or the TNO random dot test, where the eyes are dissociated by red–green spectacles.

57 **A** = False **B** = True **C** = False **D** = True **E** = True

Tinted glasses can be used for comfort (e.g. in tropical countries) or for therapeutic purposes (for example, in albinism and aniridia when photophobia is marked). There is no reason why myopic correction should not be tinted if indicated.

58 **A** = False **B** = False **C** = True **D** = True **E** = True

Pseudomonas infection is more common in extended high water content soft-lens wearers. It is a sight threatening condition and must be taken seriously. Steroids instillation predisposes the cornea to this infection. The symptoms include severe redness, pain, photophobia and purulent discharge. Swabs and corneal scrapes must be taken immediately, and microscopy and culture performed. Antibiotics (for example gentamicin) should be given subconjunctivally and topically in the first instance. If endophthalmitis develops, surgical intervention may be required to deliver antibiotics to the vitreous.

59 **A** = True **B** = True **C** = False **D** = False **E** = True

The AC/A ratio is the amount of convergence (in prism dioptres) for every dioptre change in accommodation. The value is fairly constant for a particular person at a particular time. It varies amongst individuals. It is not altered by correction of refractive error or by orthoptic exercises. A high value may cause excessive convergence when viewing near objects (when accommodation is maximum). Conversely, a low value may give rise to divergent squint on near vision due to the convergence deficit.

60 **A** = True **B** = True **C** = False **D** = True **E** = False

Contact lenses have less prismatic effect than spectacles in all kinds of anisometropia, as prismatic power = power × decentration, and decentration is less if contact lenses are used.

In axial anisometropia, the image size at the retina is the same as emmetropia (i.e. relative spectacle magnification is 1) if the lens is at the anterior focal plane of the eye. Hence, the spectacles are more likely to achieve this. In refractive anisometropia, the image size at the retina approaches that of emmetropia the closer the lens is to the nodal point of the eye. Hence, a contact lens is more likely to cause less aniseikonia.

In practice, however, anisometropia is likely to be a mixture of axial and refractive anisometropia. Hence, a trial of contact lenses is justified in all cases of anisometropia.

Use of contact lenses does not increase the monocular visual acuities, although the binocular visual acuity *may* be increased.

Test 4
Questions

1 Examples of electromagnetic waves include

A orange rays of light
B radio waves
C microwaves
D X-rays
E sound waves

2 Which of the following involve the principle of interference of light?

A Antireflective coatings of lenses.
B Use of a blue filter.
C Holography.
D Use of a pinhole.
E The Duochrome test.

3 In the law of reflection for a ray of light travelling from one medium (A) to the surface of another (B)

A the incident angle is the angle between the incident ray and the surface between the two media
B the reflected angle is the angle between the reflected light and the surface between the two media
C the normal is at 90° to the surface between the two media
D the angle of incidence is equal to the angle of reflection
E the incident ray, the reflected ray and the normal are in the same plane

4 A 2 cm object is placed 50 cm in front of a convex mirror with a radius of curvature of 80 cm.

A The image is inverted.
B The image is real.
C The image is behind the mirror.
D The image is 2 m from the mirror.
E The image is 8 cm long.

5 In the visual spectrum of light

A the longest wavelength is the red light
B the shortest wavelength is the violet light
C the frequency of red light is higher than violet light
D the wavelength of violet light is about 720 nm
E the frequency of red light is less than that of green light

6 Scattering of light

A occurs when light travels through a triangular glass prism
B occurs as a result of refraction of light through particles
C affects light of all wavelengths equally if the particles are large in relation to the wavelength of light
D affects light of short wavelengths more than light of long wavelengths if the particles are small
E explains why the sky looks blue

7 The following statements about astigmatism is/are correct.

A It causes eye-strain more often than hypermetropia.
B Regular astigmatism may be congenital.
C Regular astigmatism may occur following cataract surgery.
D Regular astigmatism may follow trauma.
E Corneal astigmatism may be partially corrected by lenticular astigmatism.

8 In the conoid of Sturm produced by a lens system of $+5DS/+5DC \times 30°$

A the interval of Sturm is 20 cm
B the circle of least diffusion is 15 cm from the lens
C the location of the circle of least diffusion is unchanged by the addition of +1DS lens
D the location of the circle of least diffusion is unchanged by the addition of $+2DS/-4DC \times 120°$
E the location of the circle of least diffusion is unchanged by the addition of $-1DS/+2DC \times 30°$

9 The refractive power

A of a lens with a focal length of 2 m is greater than one with a focal length of 1 m
B of a lens of 5D is greater than one of 3D
C of a convex lens with focal length of 3 metres is 3D
D of a convex lens with focal length of 2 metres is 0.5D
E of a convex lens with focal length of 5 cm is 2D

10 Which of the following statements regarding the phenomenon of fluorescence is/are correct?

A It can be demonstrated by directing blue light on to the chemical fluorescein.
B It is caused by the excited electron decaying to the ground state.
C The emitted light is always of a higher wavelength than the absorbed light.
D The absorbed photon is more energetic than the emitted photon.
E The emitted light is always green light.

11 A hypermetropic subject requires a contact lens of +8D.

A A spectacle lens of +6.5D would result in under-correction if the back vertex distance is 15 mm.
B A spectacle lens of +8D results in under-correction if the back vertex distance is 15 mm.
C A spectacle lens of +9.25D results in over-correction if the back vertex distance is 15 mm.
D A spectacle lens of +9.75D results in over-correction if the back vertex distance is 10 mm.
E A spectacle lens of +8.75 D results in under-correction if the back vertex distance is 10 mm.

12 A subject has a far point of about 50 cm and a near point of 8 cm.

A A concave lens of 5D should be prescribed.
B The refractive error is about 2D myopia.
C The amplitude of accommodation is 10.5D.
D The range of accommodation is increased if a –1D concave lens is used.
E The range of accommodation is about 42 cm.

13 Which of the following statements about simple myopia is/are correct?

A Degeneration of the fundus occurs.
B Retinal detachments are common complications.
C It does not progress beyond adolescence.
D It frequently exceeds 9D.
E It may be affected by severe illness.

14 Hypermetropia

A is most commonly caused by abnormal shortness in the length of the eye
B is commonly caused by abnormal curvature of the cornea
C usually causes reduced visual acuity in children
D may cause 'eye-strain' in children
E may cause convergent squint in children

15 A man has a far point of 10 m and a near point of 10 cm.

A He is about 4D myopic.
B His range of accommodation is more than 9 m.
C His amplitude of accommodation is more than 9D.
D His amplitude of accommodation remains unchanged when fully corrected.
E He is likely to be about 40 years of age.

16 The total amount of accommodation required to focus on an object 25 cm away

A in a 1D myopic subject is 2.5D
B in an emmetropic subject is 4D
C in a +4D hypermetropic subject is 0D
D in a 4 D myopic subject is 0D
E in a 1D hypermetropic subject is 3D

17 In hypermetropia

A the abnormal shape of the eye is usually confined to the post-equatorial segment of the eye
B degeneration of the choroid is common
C fundus appearance may simulate optic papillitis
D it occurs more commonly in females than males
E there is an increased risk of angle closure glaucoma

18 In the prescription of spectacles for presbyopic subjects

A the near correction can be predicted by published tables or graphs according to the subject's age
B a +5D lens should be prescribed for subjects with zero amplitude of accommodation
C the correction that gives the best near visual acuity should be ordered for every patient
D it is better to over-correct rather than under-correct near-correction
E a prism may be needed for high near-correction

19 An achromatic lens

A has less chromatic aberration compared with an ordinary lens
B can be made by cementing together a positive and a negative lens of the same dispersion
C can be made by cementing together two positive lenses of the same dispersion
D can be made by cementing together a positive and a negative lens of different dispersion
E can be made by cementing together two negative lenses with different dispersion

20 The radii of curvature of a meniscus lens are +5 cm (convex) and –12.5 cm (concave) Assuming that the refractive index of the lens is 1.5

A the refractive power of the lens is unchanged when placed in water
B the refractive power at the convex surface is weaker than 7D
C the refractive power at the concave surface is stronger than –3D
D the refractive power at the concave surface is weaker than –5D
E the refractive power of the lens is +1D

21 Ametropia

A includes myopia
B includes hypermetropia
C includes emmetropia
D includes astigmatism
E always occurs in both eyes

22 The advantages of intra-ocular lens implantation in aphakia include

A binocular single vision
B reduced risk of corneal dystrophy
C reduced risk of cystoid macular oedema
D certainty that refractive error is fully corrected
E minimal aniseikonia

23 A myopic subject requires a contact lens of –13D.

A A spectacle lens of –14.25D results in under-correction if the back vertex distance is 15 mm.
B A spectacle lens of –15.25D results in over-correction if the back vertex distance is 15 mm.
C A spectacle lens of –16.75D results in over-correction if the back vertex distance is 15 mm.
D A spectacle lens of –14D results in over-correction if the back vertex distance is 10 mm.
E A spectacle lens of –16D results in over-correction if the back vertex distance is 10 mm.

24 Which of the following statements regarding measurement of visual acuity is/are correct?

A An acuity of 6/5 implies that the chart is 5 m away.
B An acuity of 6/12 implies that the eye is able to resolve the constituent part of that letter to 2 minutes.
C The visual acuity depends on the level of illumination of the letters.
D An acuity of 6/18 implies that the eye is able to resolve the constituent part of that letter to 0.3 minutes.
E The gradation in size of letters in the Snellen chart follows a simple arithmetic progression.

25 The cross-cylinder

A can be used even if the approximate refractive error is not known
B can be used to refine the cylindrical power
C can be used to refine the spherical power
D can be used to refine the cylindrical axis
E is equivalent to a combinaation of plus and negative cylinders of equal power

26 Subjective methods in refraction

A should precede objective methods
B are means of verifying the refractive error estimated objectively
C include the Duochrome test
D depend on the observational ability of the subject for its effectiveness
E include the use of cross-cylinder

27 Spherical aberration

A is related to the prismatic effect of the lens
B affects the central part of the lens more than the periphery
C affects plano-concave lenses more than biconcave lenses
D may be reduced by occluding the periphery of the lenses
E may be reduced by using a doublet

28 Which of the following statements about retinoscopy reflex is/are correct?

A The reflex is brightest when the refractive error is nearly neutralised.
B The reflex is largest when the refractive error is large.
C The 'break phenomenon' occurs when the cylindrical lens is of the wrong power.
D The 'break phenomenon' occurs when the cylindrical lens is placed at the wrong axis.
E The speed of movement of reflex is faster when the refractive error is large.

29 Placido's disc

A is useful in examining crystalline lens
B contains a convex viewing lens
C is best used when the subject's eye is well illuminated
D is useful in detecting keratoconus
E has concentric red and green rings

30 Methods used to refine cylindrical correction include

A Friedenwald's chart
B the cross-cylinder
C Lancaster's chart
D Wallace's chart
E Worth's four-dot test

31 Which of the following statements is/are true regarding the use of cycloplegics in refraction?

A It is usually advisable in young children.
B It should be used for all subjects.
C Atropine is always the drug of choice.
D It may precipitate angle closure glaucoma.
E It may yield a result different from the true refractive state.

32 In retinoscopy

A an increase in brightness in the centre compared with the periphery can be caused by spherical aberration
B an increase in brightness in the periphery compared with the centre can be caused by spherical aberration
C problems with spherical aberration may increase with a small pupil
D if the reversal points are different in the central and peripheral area of the pupil, the average of the two points should be recorded
E a scissors-shadow usually only appears near the points of reversal

33 Which of the following statements about the Maddox rod is/are true?

A It is a series of weak convex spherical lenses mounted side by side in a trial lens.
B The patient views a light source 3 m away.
C It is traditionally placed in front of the right eye.
D It is useful in assessing both horizontal and vertical muscle imbalance.
E The line is seen to the right of the dot in esotropia.

34 Which of the following statements is/are correct concerning the use of a direct ophthalmoscope (with a plane mirror) on a hypermetropic fundus compared with a myopic fundus?

A The field of view is larger.
B The optic disc appears smaller.
C More lenses of negative power are required.
D The magnification is smaller.
E It is preferable to keep the subject's spectacles on.

35 In the use of an indirect ophthalmoscope

A the image formed is real and inverted
B the image formed is between the lens and the subject
C the image can be better seen by moving closer to the subject
D the cornea may act as a mirror causing a troublesome reflex
E the surface of the lens facing the subject may act as a mirror and causes a troublesome reflex

36 The Porro prism

A gives an inverted image
B makes use of total internal reflection
C is used in corneal microscopy
D can be fitted onto spectacle lenses
E allows the tube length in some optical instruments to be shortened

37 Oblique astigmatism in spectacle lens

A occurs when a ray of light enters the lens obliquely
B may occur if the lens is tilted
C may be reduced by bending the lens so that the outer surface is more convex
D may be utilised in subjects with sphero-cylindrical refractive error by prescription of a spherical lens with a tilt
E is less pronounced the higher the power of the lens

38 The Geneva lens measure

A measures the radius of curvature of a lens
B gives a reading of the focal length of the lens
C gives a reliable and valid reading of the overall power of the lens irrespective of the refractive index of the lens
D gives a reliable and valid reading of the overall power of the lens irrespective of the thickness of the lens
E gives a reading on the dial of the instrument

39 The three-mirror contact lens

A is made of material of the same refractive index as the eye
B can be used to view the posterior vitreous
C must be used with the slip-lamp
D can be used to view the angle of the anterior chamber
E should be applied to the eye only after application of local anaesthetic drops

40 Regarding testing of visual acuity in children

A the Sheridan–Gardner test type includes the letter A
B the Sheridan–Gardner test type includes the letter B
C the reliability depends on their interest in the test
D pictures can be made to conform to Snellen's principles
E the child points to a hand drawn on a book in the Sjögren's Hand Test

41 Which of the following statements about Worth's four-dot test is/are true?

A It can be used to test for squint.
B It can be used to test for suppression.
C If the subject sees four lights with a manifest squint, abnormal retinal correspondence is present.
D If the subject sees three green lights, normal fusion is present.
E If the subject sees five lights, diplopia is present.

42 Which of the following statements about the Varilux lens is/are true?

A The upper portion is for distant vision.
B The middle portion is of constant refractive power.
C The lower portion is of progressively increasing refractive power.
D The lateral parts of the lenses may cause significant aberration.
E It is less satisfactory if there is significant oblique astigmatism.

43 Which of the following statements about the use of a trial frame is/are correct?

A It should not be lighter than 300 g.
B It contains cells so arranged that the trial lenses are separated by at least 0.8 cm from each other.
C The spherical lens should be placed closest to the eye.
D a centring device can be used to ensure accurate centring.
E It usually accommodates a cylindrical lens with a handle.

44 A Galilean telescope system designed to view distant objects consists of two lenses of −10D and a +5D.

A The +5D lens is the eyepiece.
B The image is erect.
C The distance between the lenses is less than 15 cm.
D The system is more compact than an astronomical telescope system.
E The angular magnification is less than 5 times.

45 Two convex lenses are employed in the following

A A compound microscope.
B An astronomical telescope.
C A Galilean telescope.
D Hruby lens.
E A camera.

46 In the fitting of contact lenses for astigmatism

A the cylinder axis in the toric anterior surface can be oriented by use of a prism ballast lens
B the orientation of the lens is unimportant
C the cylinder axis can be orientated by use of truncation
D the use of a prism ballast lens is ruled out by its strong prismatic effect
E the cylinder axis can be oriented by judging the visual effect when the lens is rotated

47 Use of contact lenses in the correction of keratoconus

A is contra-indicated
B reduces the progression of the intrinsic corneal pathology
C tends to create a large area of central touch if ordinary lenses are used
D eliminates any need for corneal grafting
E is often less satisfactory than correction by spectacles

48 In indirect ophthalmoscopy

A the image is real
B the image is erect
C the image is magnified
D the amount of magnification/diminution is independent of the distance between the lens and the eye
E the image with a +20D lens is larger than with a +13D lens

49 The following general principles regarding fitting of contact lenses is/are true.

A The lens is best sealed onto the eye.
B The optical axis of the lens should coincide with the visual axis.
C The peripheral fit should prevent any tear movement.
D A maximum of dynamic fit should be aimed for.
E The fit should be reasonably stable.

50 The following is/are true concerning correction of astigmatism by contact lenses.

A A spherical hard lens is likely to correct all posterior corneal surface astigmatism.
B A spherical hard lens is likely to correct all lenticular astigmatism.
C Residual astigmatism is the astigmatism left uncorrected by a spherical contact lens.
D A toric back surface of a hard lens may induce astigmatism.
E A thin soft lens is likely to correct any corneal astigmatism.

51 Assuming that the distance between the nodal point of the eye to the 1.5 mm diameter optic nerve head is 16.6 mm

A the diameter of the blind spot at 20 cm away is 1.8 cm
B the diameter of the blind spot at 25 cm away is 4.5 cm
C the diameter of the blind spot at 50 cm away is 4.5 mm
D the diameter of the blind spot at 1 m away is 9 cm
E the diameter of the blind spot at 2 m away is 1.8 cm

52 The radii of curvature of a biconvex lens are 10 cm at one surface (A) and 12.5 cm at the other surface (B). Assuming that the refractive indices of the lens and water are 1.5 and 1.33 respectively

A the refractive power of surface A in air is 5D
B the refractive power of surface A in water is less than 2D
C the refractive power of surface B in air is 5D
D the refractive power of the lens in air is 9D
E the refractive power of the lens in water is just above 3D

53 The following statements regarding accommodation and convergence is/are true.

A Accommodation can never be exercised without convergence.
B There is unlimited dissociation between accommodation and convergence.
C The hypermetropic eye has to use accommodation more than convergence.
D Convergence is totally abolished if atropine is instilled in both eyes.
E The degree of dissociation between accommodation and convergence varies amongst individuals.

54 Correction of myopia by contact lenses compared with spectacles

A has an increased field of vision
B requires presbyopic correction later
C reduces image distortion
D reduces retinal image
E reduces the effort of accommodation

55 In the prescription of prisms for paralytic squints

A vertical squints are easier to manage than horizontal squints
B it is possible to eliminate diplopia to a reasonable extent
C it is sometimes best to divide the prismatic correction between the two eyes
D a base-down prism should be used for the hypotropic eye
E unequal base-down prisms can be used if vertical squints are more marked on the down gaze than in the primary position

56 Which of the following statements regarding amblyopia is/are true?

A It may be caused by congenital ptosis.
B Visual acuity may be better when single letters rather than rows of letters are presented to the subject.
C It may be caused by uncorrected squint in early childhood.
D It may occur with anisometropia.
E There may be a large reduction in visual acuity when a neutral density filter is added.

57 The following are valid reasons for stabilising contact lenses.

A To produce prismatic effects.
B To maintain position for bifocal/multifocal lenses.
C To avoid keratometry.
D To prevent the upper lid from dislodging the lens.
E To maintain centration of a toric surface.

58 The following are generally regarded as contra-indications to intra-ocular implantation.

A Neonates.
B Severe panuveitis.
C Myopia of –20D.
D Age above 70.
E Diabetic retinopathy.

59 Purkiinje's images are contributed by

A reflection at the anterior surface of the cornea
B reflection at the posterior surface of the cornea
C reflection at the retina
D reflection at the anterior surface of the lens
E reflection at the posterior surface of the lens

60 Which of the following statements about the Maddox wing is/are true?

A It is similar in structure to the Maddox rod.
B It can be used to demonstrate squints for near vision.
C It produces dissociation of the two eyes by mechanically presenting two different parts of the field to each of the two eyes.
D The right eye sees a white arrow pointing vertically upwards and a red horizontal arrow.
E It can be used to detect near cyclophoria.

TEST 4
Answers

1 **A** = True **B** = True **C** = True **D** = True **E** = False

See also Test 1 Question 1. Electromagnetic waves are transverse waves produced by constantly changing combined electric and magnetic fields travelling at a speed of 300 million metres per second. They do not require a medium for the waves to travel. Electromagnetic waves with a higher frequency than visible light include ultraviolet waves, X-rays and gamma rays; while electromagnetic waves with a lower frequency than visible light include infrared, microwaves, radio and TV waves. Sound waves are longitudinal waves travelling at a much lower velocity. They are caused by the vibration of air, and require air as a medium for the transmission of the waves.

2 **A** = True **B** = False **C** = True **D** = False **E** = False

If two waves having the same polarisation and wavelength fall on the same point, they are said to be coherent. Coherent waves produce constant resultant amplitude at that point. When the phase difference is zero (i.e. when they are in phase), the maxima and minima of the two waves are added together to quadruple the maximum amplitude of either wave. When the waves are 180° out of phase, the maxima of one wave is added on to the minimum of the other to produce zero amplitude. This explains the multicoloured fringes seen on oil slicks.

In ophthalmology, interference is made use of in antireflection coatings. A transparent dielectric material is evaporated on to a lens of thickness of one quarter of a wavelength. The reflected light from the lens must travel twice more through the dielectric material than the reflected light from the dielectric material. Therefore, the reflection is 180° out of phase, resulting in cancellation of the two sets of waves. Hence, there will be almost 100% transmission and no reflection.

In holography, interference is made use of by defocusing and splitting light from the laser into two beams, the reference and the object beams. The reference beam is directly shone upon a photographic plate. The object beam is directed on to the object and reflected by the object onto the same photographic plate as the reference beam. If the film is developed and another reference beam is shone through the developed negative, the second reference beam will interact with the stored interference pattern to exactly reproduce the object beam, and give a three-dimensional image.

3 **A** = False **B** = False **C** = True **D** = True **E** = True

The normal is the line perpendicular to the reflecting surface. The angle of incidence is the angle between the incident ray and the normal. The angle of reflection is the angle between the reflected ray and the normal. The laws of reflection state that

1. the incident ray, the normal ray and the reflected ray are all in the same plane;
2. the angle of incidence equals the angle of reflection.

See Fig. 4.1.

4 **A** = False **B** = False **C** = True **D** = False **E** = False

Using the ray diagram and the two basic rules

1. a ray passing through the centre of curvature is reflected along itself;
2. rays parallel to the axis are reflected passing through the focal point.

Alternatively, formulae can be used to find the position of the image. See Fig. 4.2.

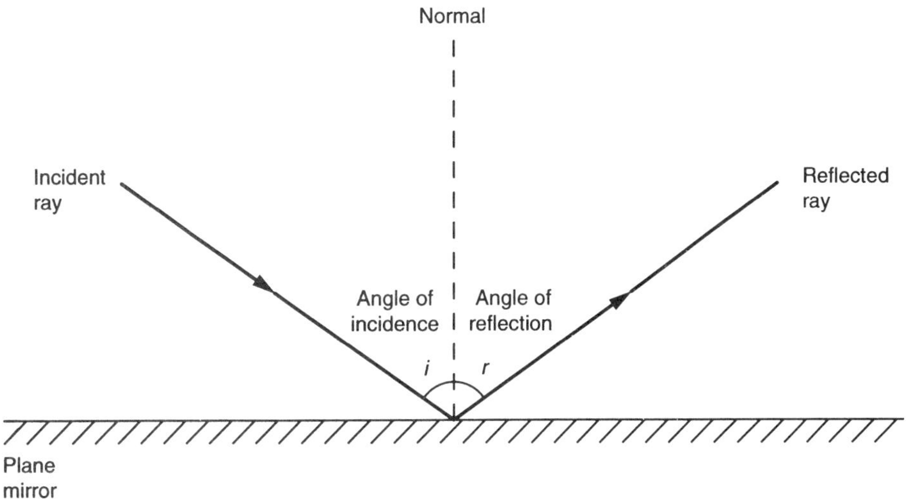

Figure 4.1 The law of reflection – plane mirror.

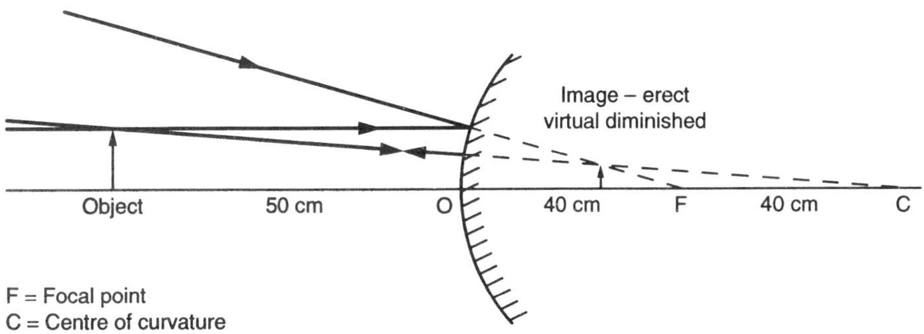

Figure 4.2 Ray diagram showing reflection in a convex mirror.

5 **A** = True **B** = True **C** = False **D** = False **E** = True

See also Test 1 Question 1. The visible spectrum of light ranges from 400 nm (blue end) and 760 nm (red end). Since the velocity of light is constant, light with a shorter wavelength has the higher frequency.

6 **A** = False **B** = False **C** = True **D** = True **E** = True

Scattering occurs when light is **reflected** off by particles in the medium in which it travels. Hence, it does not occur when travelling through a vacuum. Light of all wavelengths is equally affected if the particles are large in relation to the wavelength of light. If the particles are small, light with shorter wavelengths is more likely to be reflected. Hence, blue light is more likely to be scattered than red light, and this explains why the sky looks blue.

7 **A** = True **B** = True **C** = True **D** = True **E** = True

Astigmatism causes asthenopia (eye-strain) more often than myopia or hypermetropia. Dull ache around the eye and headaches are common. Regular astigmatism is commonly congenital. However, it may be caused by trauma or surgical procedures, especially cataract extraction. Astigmatism may be caused by abnormality in the cornea or the lens. The astigmatism caused by abnormality in the cornea may be additive to that caused by abnormality in the lens if they are of the same sign and direction, or they may partially correct each other.

8 **A** = False **B** = False **C** = False **D** = True **E** = True

The powers at the focal lines are 5D and 10D. They are 1/5 m = 20 cm and 1/10 m = 10 cm away from the lens. Hence, the conoid of Sturm is (20 – 10) cm = 10 cm.

At the circle of least diffusion, it has a dioptrical power of (5 + 10)/2 = 7.5D. Hence, it is (1/7.5) m = 13.3 cm away from the lens.

Addition of +1DS would cause the power at the focal lines to increase to 6D and 11D. At the circle of least diffusion, it would have a dioptrical power of (6 + 11)/2 = 8.5D. The circle of least diffusion would be (1/8.5) m = 11.8 cm away from the lens.

Addition of –1DC × 120° would increase the power at the focal lines to be 6D and 10D. Hence, the power at the circle of least diffusion would be (6 + 10)/2 = 8D. It would be 12.5 cm away from the lens. In general, addition of a positive cylindrical lens would move that focal line closer to the lens, and addition of a negative cylindrical lens would move that focal line away from the lens.

Addition of –1DS/+2DC × 30° would change the power at the focal lines to be 4D and 11D. At the circle of least diffusion, it would have a power of (4 + 11)/2 = 7.5D. Hence, it is at the same distance of 13.3 cm away from the lens.

Addition of +2DS/–4DC × 120°is the equivalent of addition of –2DS/+4DC × 30°. Using similar calculations, it can be shown that the position of the circle of least diffusion is unchanged.

9 **A** = False **B** = True **C** = False **D** = True **E** = False

Focal length is defined as the distance between the optical centre of the lens and its principal focus. The refractive power of a lens is measured in D. A lens with focal length × metres has a refractive power of $1/x$ D. The shorter the focal length, the greater the refractive power.

Hence, a lens with focal length of 2 m has a refractive power of 1/2 = 0.5D, while that with a focal length of 1 m has a refractive power of 1D.

A convex lens with a focal length of 3 m has a refractive power of 1/3 = 0.33D.

A convex lens with a focal length of 2 m has a refractive power of 1/2 = 0.5D. A convex lens with a focal length 5 cm (0.05 m) has a refractive power of 1/0.05 = 20D.

10 **A** = True **B** = False **C** = True **D** = True **E** = False

If light is directed to and absorbed by a material, the energy of the photon may cause an electron of the material to be raised to an excited state. Fluorescence refers to the decay of the electron from an excited state to a state higher than the ground state. A familiar example in ophthalmology is the direction of blue light onto chemical fluorescein, and a green light (lower frequency, hence less energy than the blue light) is emitted. However, there are other substances which absorb and emit light of different frequencies. The emitted light is always of less energy (hence lower frequency and higher wavelength) than the absorbed light.

11 **A** = True **B** = False **C** = True **D** = True **E** = False

Applying the formula

$$F_d = \frac{F}{1 + dF}$$

where F_d = refractive power of the correcting lens placed d metres away from the eye (negative if moved towards the eye), F = refractive power of the correcting lens of the original lens in D (+ve for convex lens, –ve for concave lens) and d = distance moved away from eye (negative if moved towards eye).

If the Back Vertex Distance (BVD) is 15 mm (0.015 m)

$$\begin{aligned} F &= 8\text{D} \\ d &= 0.015 \text{ m} \\ F_d &= 8/(1 + 0.015 \times 8) \\ &= 7.14\text{D} \end{aligned}$$

If the BVD is 10 mm

$$\begin{aligned} F &= 8\text{D} \\ d &= 0.01 \text{ m} \\ F_d &= 8/(1 + 0.01 \times 8) \\ &= 7.41\text{D} \end{aligned}$$

12 **A** = False **B** = True **C** = True **D** = True **E** = True

Refractive error refers to the static refraction of the eye (i.e. when the eye is not accommodated). The far point is the furthest point objects can be seen without blurring. In emmetropia, this is optical infinity. If the far point is 50 cm (0.5 m), the eye has an excess of 1/0.5D = 2D refractive power. Hence, the subject is 2D myopic.

The amplitude of accommodation is the difference in refractive power of the eye when the eye is fully accommodated and at rest. The near point is the closest point a subject can see small objects without blurring. The eye would be fully accommodated. The near point of the subject is 8 cm. Hence, with full accommodation, the refractive power of the eye is 1/0.08 D = 12.5D more than the emmetropic eye. Hence, the amplitude of accommodation is 12.5D – 2D = 10.5D.

The range of accommodation is the distance between the far and the near point. In the subject, the range of accommodation = 50 cm – 8 cm = 42 cm.

If a –1D concave lens is used, the subject would be 1D myopic, and the far point is 1 m. Since the amplitude of accommodation is 10.5D, the refractive power would be 11.5D more than the emmetropic eye when fully accommodated. Hence, the near point is 1/11.5 = 0.09 m = 9 cm. The range of accommodation is (100 – 9) cm = 91 cm.

13 **A** = False **B** = False **C** = True **D** = False **E** = True

Simple myopia can be thought of as normal physiological variations of axial length and curvature of the cornea. It does not progress beyond adolescent, and it seldom exceeds 6D. Peripheral retinal degeneration may occur from middle age, but degenerative changes in the fundus do not occur. In severe illness, the sclera may become stretched. The axial length is increased, and the myopia is made worse.

14 **A** = True **B** = False **C** = False **D** = True **E** = True

Hypermetropia may be caused by axial hypermetropia, refractive hypermetropia or index hypermetropia. Axial hypermetropia (caused by abnormal shortness of the eye) is the most common. Hypermetropia rarely exceeds about 7D. Children have a large amplitude of accommodation so that any hypermetropia is likely to be compensated by accommodation. Hence, blurring of vision does not occur. However, they may complain of 'eye-strain' due to persistent tension in the ciliary muscle. This is especially noticeable after close work. The symptoms may include aching red watery eyes, and headaches. Convergent squint may occur in children especially those with a high AC/A ratio.

15 **A** = False **B** = True **C** = True **D** = True **E** = False

See also Test 4 Question 12. His far point is 10 m. Hence, he is 1/10D = 0.1D myopic. His range of accommodation is the distance between the far point and the near point = 10 m – 10 cm = 9.9 m. His amplitude of accommodation is the difference in refractive power of the eye when fully accommodated and at rest.

His refractive power when fully accommodated is 1/0.1 = 10D. Hence, his amplitude of accommodation = (10 – 0.1)D = 9.9D. A man of 40 years of age would be expected to have an amplitude of accommodation of about 4.5D. The man with an amplitude of accommodation of 9.9D is likely to be around 20 years old. The amplitude of accommodation does not vary with spectacle lenses used.

16 **A** = False **B** = True **C** = False **D** = True **E** = False

The total accommodation required to focus on an object 25 cm away in an emmetropic subject is (1/0.25) = 4D.

In a 1D myopic person, there is an excess of 1D in the eye. Hence, only (4 – 1) = 3D is required.

In a +4D hypermetropic person, 4D is used to overcome the hypermetropia. Hence (4 + 4) = 8D of accommodation is required.

In a –4D myopic person, there is an excess of 4D in the eye, and the far point is 1/4 = 0.25 m from the eye. Hence, no accommodation is required.

Similarly, in a 1D hypermetropic person, the amount of accommodation required is (4 + 1) = 5D.

17 **A** = False **B** = False **C** = True **D** = False **E** = True

Unlike myopia, the abnormality in the size and shape of the eye is not confined to the posterior segment of the eye. The diameter of the cornea may be reduced, and the anterior chamber of the eye is usually shallow. Hence, there is an increased risk of angle closure glaucoma. There is no difference in incidence between male and female. The optic disc may simulate optic papillitis. Unlike myopia, degeneration of choroid and retina does not occur.

18 **A** = False **B** = False **C** = False **D** = False **E** = True

The near-correction should be determined for each patient individually for two reasons. First, the decline of the amplitude of accommodation varies amongst individuals. Secondly, each subject has their own activities which require vision at different distances. Comfort when reading and adjustment by individual subjects are more important than the recorded near visual acuity in the prescription for near-correction.

The maximum near-correction that should generally be prescribed is probably about 3–3.5D. This would allow for a reading distance of about 33 cm. Correction above this is generally not well tolerated. It is generally better to under-correct than over-correct. If high correction is prescribed, a prism should be considered to help with deficiency in convergence.

19 **A** = True **B** = False **C** = False **D** = True **E** = False

Achromatic lenses are designed to eliminate chromatic aberration. This is achieved by cementing together a positive and negative lens of different dispersions, so that the refractive powers for light at the two ends of the spectrum are equal to the desired refractive power. Cementing together lenses of the same dispersion would not eliminate chromatic aberration.

20 **A** = False **B** = False **C** = True **D** = True **E** = False

Applying the formula

$$\text{power of surface (in D)} = (n_1 - n_2)/r$$

where n_1 = refractive index of lens, n_2 = refractive index of air and r = radius of curvature of lens in m (+ve if convex, –ve if concave)

The refractive power at the convex surface = (1.5 – 1)/0.05 = 10D.
The refractive power at the concave surface = (1.5 – 1)/–0.125 = –4D.

Hence, refractive power of the lens = 10 – 4 = 6D.

If the lens is placed in water, the refractive index of the lens relative to water would be 1.5/1.33 = 1.1.

21 **A** = True **B** = True **C** = False **D** = True **E** = False

Ametropia means errors of refraction (literally, not according to measure). It includes myopia, hypermetropia and astigmatism. All these conditions can occur in one eye only.

22 **A** = True **B** = False **C** = False **D** = False **E** = True

The advantages of intra-ocular lens implantation are convenience, good binocular single vision, improved peripheral field, and minimal difference in retinal image size between the two eyes (i.e. minimal aniseikonia). There is a slight risk of corneal dystrophy and cystoid macular oedema, in addition to pupil block glaucoma, postoperative astigmatism (which may be minimised by removal or insertion of stitches), post-operative uveitis, posterior capsule rupture requiring vitrectomy, and intra-ocular lens dislocation.

23 **A** = True **B** = False **C** = True **D** = False **E** = True

Applying the formula

$$F_d = \frac{F}{1 + dF}$$

where F_d = refractive power of the correcting lens placed d metres away from the eye (negative if moved towards the eye), F = refractive power of the correcting lens of the original lens in D (+ve for convex lens, –ve for concave lens) and d = distance moved away from eye (negative if moved towards eye).

If the Back Vertex Distance (BVD) is 15 mm (0.015 m)

$$\begin{aligned} F &= -13\text{D} \\ d &= 0.015\text{ m} \\ F_d &= -13/(1 + 0.015 \times -13) \\ &= -16.15\text{D} \end{aligned}$$

If the BVD is 10 mm (0.01 m)

$$F = -13\text{D}$$
$$d = 0.01 \text{ m}$$
$$F_d = -13/(1 + 0.01 \times -13)$$
$$= -14.94\text{D}$$

24 **A** = False **B** = True **C** = True **D** = False **E** = False

A visual acuity of 6/5 implies that a subject reads at a distance of 6 m what a 'normal' subject reads at a distance of 5 m. A 'normal' subject is taken to be able to resolve the constituent parts of a letter to 1 minute. Hence, a visual acuity of 6/12 and 6/18 imply that the subject has a resolution of 2 minutes and 3 minutes respectively. The gradations in size in a Snellen charts are 6/60, 6/36, 6/24, 6/18, 6/9, 6/6, 6/5, 6/4, which do not follow a simple arithmetic progression. The measurement of the visual acuity by Snellen charts depends on the illumination of the letters. It also depends on how close the letters are. This is especially important for children who have amblyopia or eccentric fixation.

25 **A** = False **B** = True **C** = False **D** = True **E** = True

See also Test 1 Question 25. A cross-cylinder is equivalent to two cylinders of equal but opposite power with axes at right angles to each other. It can used to refine both the cylindrical power and the cylindrical axis. The refractive error must be roughly corrected when the cross-cylinder is used, so that the circle of least diffusion is close to the retina. Addition of cross-cylinders does not affect the position of the circle of least diffusion.

26 **A** = False **B** = True **C** = True **D** = True **E** = True

Subjective methods of refraction rely on the subject assessing the clarity of various targets. This depends on the cooperation, education and power of observation of the subject for the methods to be reliable and accurate. Hence, objective methods (e.g. retinoscopy, keratometry) should be performed before subjective methods are used. Subjective methods include the Duochrome test, use of a cross-cylinder and use of an astigmatic fan/block.

27 **A** = True **B** = False **C** = False **D** = True **E** = True

Also see Test 1 Question 27. Plano-concave and plano-convex lenses have less spherical aberration compared with biconcave and biconvex lenses respectively. Since the spherical aberration is greatest at the periphery, it may be reduced by occluding the periphery of the lens. A doublet consists of a main lens and a weaker lens of the opposite sign and different refractive index placed together, so that the peripheral prismatic effect of the weaker lens can neutralise that of the main lens.

28 **A** = True **B** = False **C** = False **D** = True **E** = False

The size of refractive error can be estimated by the brightness, size of reflex and speed of reflex movement in retinoscopy. The reflex is brighter and larger, and moves more rapidly when it is near to neutrality. The 'break phenomenon' occurs when the reflex seen in the pupil is not continuous with the streak outside the pupil. This occurs when the streak is directed at the wrong axis to the subject's astigmatic axis. If the cylindrical lens is of the wrong power but placed at the right axis, the reflex would move either with or in the opposite direction to the retinoscope. However, the 'break phenomenon' would not occur.

29 **A** = False **B** = True **C** = True **D** = True **E** = False

The Placido's disc contains a plane circular disc with alternate white and black concentric rings, and a convex viewing lens in the centre of the disc. Light should be directed at the subject's eye, and the examiner looks through the central convex lens and observes the reflection of the image of the disc from the subject's cornea. If the subject's cornea is regular, the image of the disc would be circular. Any irregularity of the subject's cornea would cause distortion of the image. It is not useful in studying the crystalline lens of the subject.

30 **A** = True **B** = True **C** = True **D** = True **E** = False

Techniques used to refine cylindrical correction include the cross-cylinder and astigmatic figures. Astigmatic lines in the same direction as the line of greatest refractive error would be the most clearly defined. There are many types of astigmatic figures. They can be arranged in a fan shape as in Landolt's chart, or in a circular form as in Lancaster's chart, or Wallace chart. In Wallace chart, the lines are duplicated. In Friedenwald's chart, black lines with white borders set in a grey background are used. Worth's four-dot test is used to test extra-ocular muscle balance.

31 **A** = True **B** = False **C** = False **D** = True **E** = True

The use of cycloplegics is advisable in young children as they have great amplitude of accommodation and are generally unable to relax their accommodation voluntarily. It should not be used in all subjects, and certainly not in older subjects with minimal amplitude of accommodation. Cyclopentolate is often sufficient, and the effect of atropine can last for up to 2 weeks. There is always a risk of precipitating angle closure glaucoma. Anterior chamber depth and gonioscopy should ideally be performed before instillation of mydriatics and cycloplegics.

Refraction under cycloplegics may yield a refractive result different from the true refractive state for three reasons. First, the lens assumes a different shape under cycloplegia. Secondly, pupil dilatation may result in the examiner refracting the peripheral areas of the pupil, where spherical aberration may be great. Thirdly, the subject may have accommodative abnormalities (e.g. accommodative spasm) which would not be taken into account with cycloplegic refraction.

32 **A** = True **B** = True **C** = False **D** = False **E** = True

Spherical aberration usually causes light to be refracted more at the periphery than the centre. At the point of reversal for the central zone of the pupil, there will be an increase in brightness in the centre compared with the periphery. At the point of reversal for the peripheral zone of the pupil, there is an increase in brightness in the periphery compared with the centre. A small pupil generally reduces the spherical aberration by cutting down peripheral rays of light. If the reversal points are different for the central and peripheral parts of the pupil, the reversal point for the centre part should be recorded, as this is the optical zone.

Scissors-shadow is usually caused by an irregular astigmatic cornea or lens. Scissors-shadow occurs when one part of the cornea shows the phenomenon of myopia while another part shows the phenomenon of hypermetropia, causing the scissors-shadow at or near the point of reversal.

33 **A** = False **B** = False **C** = True **D** = True **E** = True

The Maddox rod consists of a series of high powered red-tinted cylindrical lenses mounted side by side on a trial lens. It is traditionally placed in front of the right eye of the subject, and at least 6 m away from the eye (optical infinity). A white point source is viewed, and an image of a red line at 90° to the axis of the cylinder is formed.

The Maddox rod therefore dissociates the two eyes. The right eye sees a red line at 90° to the axis of the cylinders, while the left eye sees the white point source. It can be used to assess horizontal or vertical muscle imbalance, as well as cyclophoria.

To assess horizontal muscle imbalance, the Maddox rod must be placed with the axes of the cylinders horizontally, and a vertical red line is seen. In general, deviation of the right eye in one direction would cause the red line to be seen to be deviated in the opposite direction. Hence, in esophoria (convergent squint), the red line is seen to the right of the white point source. In exophoria (divergent squint), the red line is seen to the left of the white point source (cross diploplia). Vertical muscle imbalance can be assessed in the same way with the axes of the cylinders vertical.

To assess cyclophoria, a second untinted Maddox rod can be added in front of the left eye, with both Maddox rods with their axes in the same direction. If the subject has cyclophoria, the red and the white lines would be seen at an angle to each other. One of the Maddox rods can then be adjusted to make the lines coincide (or be parallel if the subject has horizontal muscle balance as well). The degree of cyclophoria is equal to the angle adjustment necessary (see Fig. 4.3).

34 **A** = True **B** = True **C** = False **D** = True **E** = False

When the direct ophthalmoscope is used, one can regard the eye itself as a convex lens, and the magnification can be calculated as

$$\text{magnification} = \text{refractive power of eye}/4.$$

A myopic subject has a higher refractive power than a hypermetropic subject.

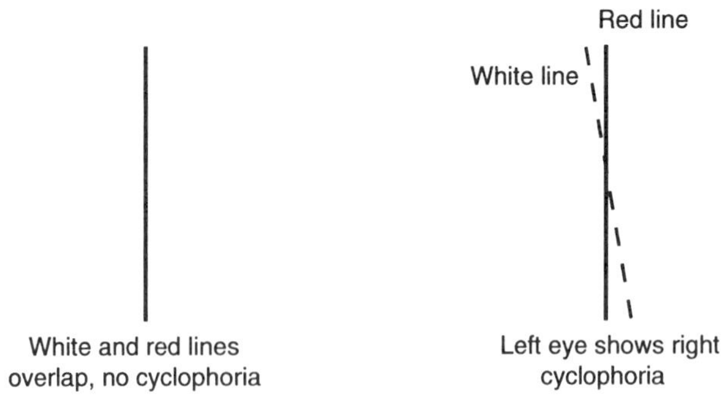

Figure 4.3 The Maddox rod.

Hence magnification is less in hypermetropic subjects than myopic subjects. The field of view is correspondingly smaller. It is preferable to keep the subject's spectacles on when viewing the fundus of a highly myopic patient.

35 **A** = True **B** = False **C** = False **D** = True **E** = True

The image formed by the indirect ophthalmoscope is real and inverted. It is formed between the lens and the examiner. Hence, there is a natural tendency to get closer to the subject, which must be resisted. By first principles, the lens is best placed at a distance from the subject's eye equal to the focal length of the lens. However, there are many reflexes which interfere with the examiner's view. First, the cornea acts as a mirror and forms a reflex of the light source which can be further magnified by the convex lens. Second, both surfaces of the lens can act as mirrors and create troublesome reflexes.

36 **A** = True **B** = True **C** = True **D** = False **E** = True

The Porro prism makes use of total internal reflection at two surfaces to give an inverted image. It is used in slit lamp microscopy between the objective and the eyepiece to shorten the tube length, and to reinvert the inverted image (see Fig. 4.4).

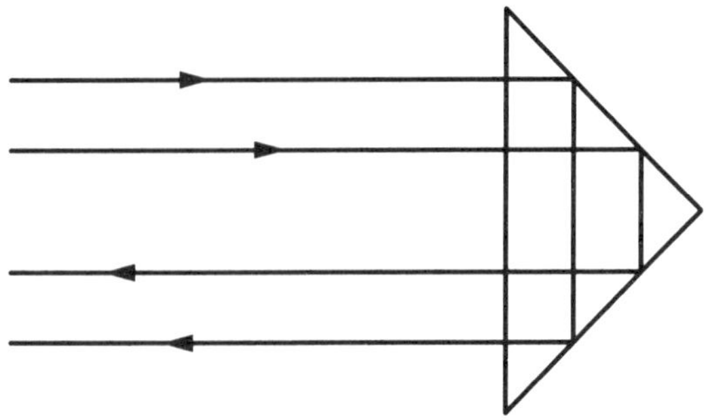

Figure 4.4 The Porro prism.

37 **A** = True **B** = True **C** = True **D** = True **E** = False

Oblique astigmatism occurs when light enters the lens obliquely. Two focal lines are formed, and two focal surfaces are traced out as the light rotates around the optical axis of the lens. The same effect is achieved by tilting the lens. It may be reduced by bending the lens so that the outer surface becomes more convex, which also reduces spherical aberration. Oblique astigmatism is more pronounced the higher the power of the lens. It may be utilised in some cases by prescribing a spherical lens with a tilt for subjects with a large spherical and a small horizontal cylindrical refractive error.

38 **A** = True **B** = False **C** = False **D** = False **E** = True

The Geneva lens measure estimates the refractive power of a lens by measuring the radius of curvature of a lens. It assumes the refractive index of the lens to be 1.52 (that of crown glass), and converts the reading into refractive power in D. The reading is displaced on the dial of the instrument. To calculate the total refractive power of a lens, readings at both surfaces of the lens are taken, and the refractive powers are added together. This assumes that the thickness of the lens is negligible.

39 **A** = False **B** = True **C** = True **D** = True **E** = True

The Goldman three-mirror contact lens is a plano-concave lens shaped like a cone. It is used to view the posterior, middle and peripheral parts of the retina, as well as the anterior chamber angle. It is made of materials of higher refractive index than the eye to destroy the air/cornea interface, and overcomes the problem of total internal reflection when light emerges from the anterior chamber angle.

The central part of the lens is used to view the central part of the fundus. There are three mirrors placed with different angles of inclination, to allow different areas of the fundus to be brought into focus. The lens will not be tolerated by the patient without prior instillation of local anaesthetic drops.

40 **A** = True **B** = False **C** = True **D** = True **E** = False

The Sheridan–Gardner test type includes seven symmetrical letters (Letters H, V, O, X, U, A, T), so that it can be used with a mirror in a 3 m room. The reliability of any test type for children depends more on the children's interest and understanding of the test than on theoretical considerations. Pictures can be made into different sizes and to conform to Snellen's principles of targets subjecting a pre-defined visual angle. In the Sjogren's Hand Test, the child points his/her own hand in the same direction as one shown on a card.

41 **A** = False **B** = True **C** = True **D** = False **E** = True

Worth's four-dot test is used to test for abnormality of fusion. If the subject sees four lights with a manifest squint, fusion appears to be present. (The image of the white light is fused.) The presence of a squint means that the subject must be fixating with the extra-foveal area of one eye. Hence, abnormal retinal correspondence is present. If three green lights are seen, all the images are perceived by the left eye. Hence, there is right eye suppression. If the subject sees five lights, both eyes are used, but the image are not fused. There must be diplopia.

42 **A** = True **B** = False **C** = False **D** = True **E** = True

The Varilux lens is an example of a multifocal lens in which the near portion has a continuous increasing power. This avoids the prismatic jump in the bifocal lens.

The Varilux lens contains three portions: the upper portion is for distant vision; the middle portion has a 12 mm segment of progressive increasing refractive power; and the lower portion has a fixed maximum refractive power for reading. Spherical aberration in the lateral parts of the lens, may give rise to image distortion. This is made worse if there is significant oblique astigmatism.

43 **A** = False **B** = False **C** = True **D** = True **E** = False

Trial frames should be light. It contains three cells for each eye which fit the trial lens securely and close to each other. There is a handle for the spherical lens, but not for the cylindrical lens, as it would be difficult to adjust the cylindrical axis.

Adjusting the trial frame so that the optical centre of the trial lens lies in the visual axis (centring of the lens) is important. It can be achieved by the centring device which consists of plane glass with a cross meeting at the centre. The subject looks straight ahead at a light, and the examiner adjusts the frame so that the corneal reflection of the centre of the cross is centred.

44 **A** = False **B** = True **C** = True **D** = True **E** = True

The +5D lens is the objective with focal length of (1/5) m = 20 cm, and the −10D lens is the eyepiece with focal length (1/−10) m = −10 cm. The distance between the lenses is (20 − 10) = 10 cm. The angular magnification equals the ratio of the focal lengths = 20/10 = 2 times.

A Galilean telescope system is considerably more compact than the astronomical telescope system. The image is erect.

45 **A** = True **B** = True **C** = False **D** = False **E** = False

The astronomical telescope is used to magnify a distant object, and it comprises two convex lenses separated by a distance equal to the sum of their focal lengths. The objective lens forms a real image of the object, while the second lens, a very strong convex lens, acts as a simple magnifier of the real image formed by the first lens (see Fig. 4.5).

The angular magnification of an astronomical telescope
= magnified angle/object angle
= focal length of objective/focal length of eyepiece

The Galilean telescope is similar to the astronomical telescope except that the eyepiece is a concave lens, and the image formed is erect. The distance between the lens equals the difference between the two focal lengths.

The compound microscope is similar to the astronomical telescope except that

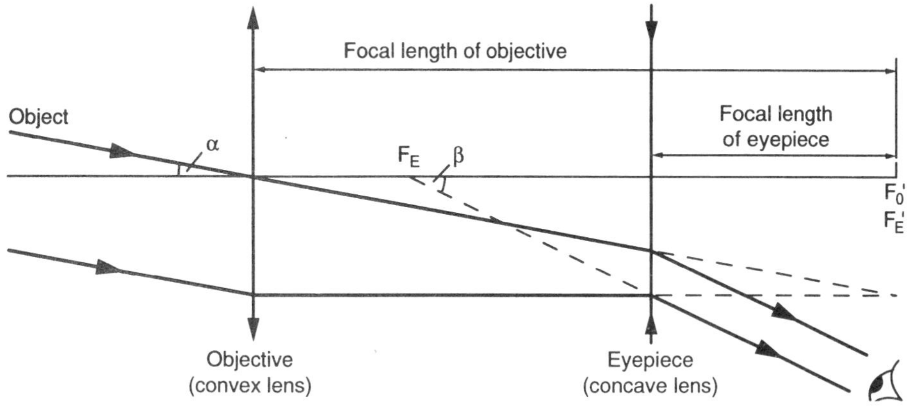

Figure 4.5 The Galilean telescope. F_e and F_e' are focal points of eyepiece, f_0 is focal point of objective. Angular magnification = β/α.

the objective lens has a very short focal length, and the object to be viewed is placed near to the focal plane of the objective lens. The eyepiece acts as a magnifier, and the total linear magnification equals the product of the magnification produced by the objective and the eyepiece.

The Hruby lens consists of one plano-concave lens of −58.6D. The camera usually consists of one convex lens, with an adjustable distance between the lens and the film.

46 **A** = True **B** = False **C** = True **D** = False **E** = False

In the prescription of a toric lens for astigmatism, it is important to align the cylindrical axis of the lens to the astigmatic axis of the eye. This can be achieved by the prism ballast lens, where a prism is added to the lower portion of the lens, which will help the lens to be positioned with the prism downwards. Prisms in the contact lens have little optical prismatic effect as the refractive index from tear to lens is less than that from air to lens. Truncation involves cutting off one edge of a lens. The truncated edge is heavier and tends to position itself downwards.

47 **A** = False **B** = False **C** = True **D** = False **E** = False

Contact lenses are useful in correcting the irregular astigmatic corneal surface in keratoconus, and are usually more satisfactory than spectacles. However, they do not reduce the progression of the intrinsic corneal pathology, and therefore do not eliminate the need for corneal grafting. Keratoconus corneas are steep centrally and flat peripherally. Hence, fitting an ordinary lens tends either to create a large area of central touch, or to create an area of central touch with bubbles in the intermediate zone.

48 **A** = True **B** = False **C** = True **D** = False **E** = False

In indirect ophthalmoscopy, a powerful convex lens (+20D or +13D) is held in front of the patient's eye, and a powerful illuminating light source is provided by a head lamp worn by the observer. A real, inverted, laterally inverted image is formed between the lens and the observer

The magnification is about 3 times when a 20D lens is used, and 5 times when a +13D lens is used.

49 **A** = False **B** = True **C** = False **D** = False **E** = True

Contact lenses should be reasonably stable (to prevent being dislodged), but the peripheral fit should allow adequate tear flow under the lens on blinking. Lenses should not be sealed onto the eye. A balance between the extreme of static fit (where lid movement does not produce lens movement) and dynamic fit (where lid movement produces lens movement in a different direction) is required. The lens should be centred (i.e. the optical axis of the lenses in alignment with the visual axis of the eye), to prevent prismatic effect.

50 **A** = False **B** = False **C** = True **D** = True **E** = False

A hard spherical contact lens corrects most of the anterior corneal surface astigmatism, but leaves the posterior corneal and lenticular astigmatism uncorrected. A toric back surface (i.e. a lens with spherical and cylindrical components on the back surface) may induce astigmatism. The astigmatism left uncorrected by a spherical hard contact lens is called the 'residual astigmatism'. A thin soft lens placed on an astigmatic cornea will not correct the astigmatism, as the lens may mould onto the cornea. A thick soft lens may partially correct the anterior corneal astigmatism.

51 **A** = True **B** = False **C** = True **D** = True **E** = False

The size of the blind spot can be calculated as the size of an object which produces the image size of the optic nerve head. From Fig. 4.6

$$\frac{\text{Diameter of blind spot size}}{\text{Distance where blind spot measured}} = \frac{\text{Diameter of optic nerve head}}{\text{Distance from nodal point to retina}}$$

At a distance of 20 cm, the diameter of blind spot = (0.15/1.66) × 20 cm = 1.8 cm
At 25 cm, the diameter of the blind spot = (0.15/1.66) × 25 cm = 2.26 cm
At 50 cm, the diameter of the blind spot = (0.15/1.66) × 50 cm = 4.5 cm
At 1 m, the diameter of the blind spot = (0.15 /1.66) × 100 cm = 9 cm
At 2 m, the diameter of the blind spot = (0.15/1.66) × 200 cm = 18 cm

52 **A** = True **B** = True **C** = False **D** = True **E** = True

Applying the formula

$$\text{Power of surface (in D)} = (n_1 - n_2)/r$$

where n_1 = refractive index of lens, n_2 = refractive index of second medium and r = radius of curvature of lens in m (+ve if convex, –ve if concave).

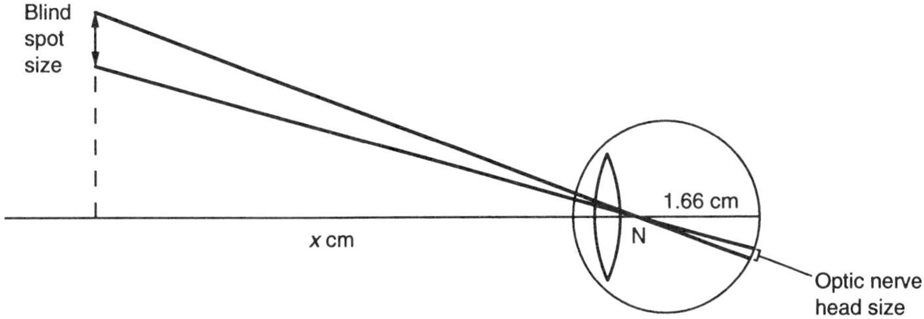

Figure 4.6 Calculation of blind spot diameter from optic nerve head diameter. N = the nodal point.

The refractive power of surface A in air = (1.5 – 1)/0.1 = 5D
The refractive power of surface A in water = (1.5 – 1.33)/0.1 = 1.7D
The refractive power of surface B in air = (1.5 – 1)/0.125 = 4D
The refractive power of surface B in water = (1.5 – 1.33)/0.125 = 1.36D

Hence, the refractive power of a lens in air = 5 + 4 = 9D.
The refractive power of a lens in water = 1.7 + 1.36D = 3.06D.

53 **A** = False **B** = False **C** = True **D** = False **E** = True

Accommodation and convergence can be separately exercised to a limited extent. This varies amongst individuals, and at different times. It appears to increase with practice. Hence, instillation of atropine into both eyes does not abolish convergence altogether. The hypermetropic eye uses accommodation to overcome the refractive error, and hence uses accommodation more than convergence. Conversely, the myopic uses convergence more than accommodation.

54 **A** = True **B** = False **C** = True **D** = False **E** = False

The advantages of correction of myopia by contact lenses include reduced prismatic effect and spherical aberration, increased image size, and increased field of vision. However, if the subject has a limited amplitude of accommodation, it may not be well tolerated. This is because there is more demand for accommodation if contact lenses are used than if spectacles are used.

55 **A** = False **B** = True **C** = True **D** = False **E** = True

Prisms can be used in paralytic squints to eliminate diplopia, achieving good binocularity in a reasonable field of vision. Horizontal squints are easier to manage than vertical squints, as vertical deviations are not well tolerated. To reduce the need for using strong prisms, it is often best to divide the prismatic correction between the two eyes. A base-up prism should be used for the hypotropic eye. Unequal base-down prisms (with the stronger one in front of the hypertropic eye) can be used if vertical squints are more marked on the down gaze than in the primary position. This will encourage the gaze of the eyes to be in areas where muscle imbalance is least.

56 **A** = True **B** = True **C** = True **D** = True **E** = False

Amblyopia may occur as a result of squint (due to continuous suppression of the deviated eye), stimulus deprivation (e.g. congenital cataract, ptosis) or anisometropia in infancy and childhood, probably up to the age of 5–7. The visual acuity may be reduced. Visual acuity may be worse when rows of letters rather than single letters are presented (crowding phenomenon). The neutral density filter test may show no reduction in visual acuity.

57 **A** = True **B** = True **C** = False **D** = False **E** = True

Valid reasons for stabilising contact lenses include:
1. producing the prismatic effect for therapeutic reasons;
2. maintaining the position for the segment for near vision for bifocal or multifocal lenses;
3. maintaining a toric lens in the correct corneal meridian.

58 **A** = True **B** = True **C** = True **D** = False **E** = False

There are considerable disagreements about contra-indications for intra-ocular implantation. Old age is certainly not a contra-indication. Most would be reluctant to insert intra-ocular lens in neonates, because the effects of intra-ocular lenses over a period of, say, 70 years, are still uncertain. Severe uveitis can be exacerbated by intra-ocular lens. Myopia above –20D does not require intra-ocular lenses. Diabetic retinopathy *per se* is not a contra-indication, though some are reluctant to insert intra-ocular lenses if treatment of retinopathy is made more difficult in the future.

59 **A** = True **B** = True **C** = False **D** = True **E** = True

The four principal Purkinje images are from the anterior and posterior surfaces of the cornea and the anterior and posterior surfaces of the crystalline lens. The first Purkinje image (from the anterior surface of the cornea) is the brightest, and the keratometer measures the radius of curvature of the anterior surface of the cornea utilising the first Purkinje image.

60 **A** = False **B** = True **C** = True **D** = True **E** = True

A Maddox wing measures the degree of squints for near vision by dissociating the two eyes (usually at about 33 cm).

The instrument is arranged so that the right eye sees a white vertical arrow and a horizontal red line, while the left eye sees both a vertical and a horizontal row of numbers. The subject is asked which numbers the white arrow and the red line point to, and the horizontal and vertical components of the squint can be determined. To determine cyclophoria, the subject is asked to move the red arrow so that it is in the same direction as the horizontal row of numbers. The angle of cyclophoria can then be measured by the angle between the red line and the horizontal row of numbers.

INDEX

Topic	Test	No.	Page No.
Basic sciences			
Accommodation			
—mechanisms	2	16	50
	3	16	77
Blind spot diameter			
calculation	4	51	113
Diffraction	1	3	15
	2	3	47
Dispersion	3	2	75
Electromagnetic waves	4	1	104
Eye—			
emmetropic	3	5	75
	2	5	48
reduced	1	5	16
Fluorescence	4	10	105
Focal length—refractive			
error calculation	3	9	76
	4	9	105
Interference	4	2	104
Lasers	2	6	48
	3	6	76
Lens			
convexo-concave	2	10	49
concave and convex	3	10	76
Light	3	1	75
Mirror—concave	3	4	75
convex	4	4	104
Parallel glass block	1	4	15
Penetration of different			
wavelengths of light	2	2	47
Polarisation	3	3	75
Presbyopia	2	18	50
Prismatic power—			
measurement	3	37	82
	2	36	54
Prisms	2	4	47
Risley	3	36	82
Porro	4	36	110
Radius of curvature—			
calculation of power			
	2	20	51
	3	20	78
	4	20	107
	4	52	113

Topic	Test	No.	Page No.
Reflection—law of	4	3	104
Refractive Index	1	2	15
Scattering	4	6	105
Spectrum of light	1	1	15
	4	5	104
	2	1	47
Total internal reflection	1	21	18
Clinical refraction			
Accommodation	1	18	18
	1	16	18
	1	58	26
	2	19	50
	4	16	107
—and convergence	4	53	114
—abnormalities	2	60	58
—Relative	1	53	25
	2	53	57
AC/A ratio	1	8	16
	3	59	86
Amblyopia	4	56	114
Ametropia	4	21	108
Anisometropia	3	23	79
Angle kappa	3	55	85
Aphakia	3	22	79
Astigmatism	1	10	16
	1	15	17
	2	15	50
	3	7	76
	4	7	105
Asthenopia	1	56	25
Binocular vision	3	56	86
Blindness (definition)	3	48	84
Conoid of Sturm	2	7	50
	2	8	48
	3	8	76
	4	8	105
	1	7	16
Convergence	3	53	85
Cycloplegics	2	29	52
	4	31	109
Far-point near-point	1	12	17
	1	19	18
	2	12	49
	3	12	77
	4	12	106
	4	15	106
Fusional reserves	2	55	57
Hypermetropia	1	14	17
	2	14	49

Topic	Test	No.	Page No.
	3	14	77
	4	14	106
	4	17	107
Maddox rod	4	33	110
	1	39	22
Maddox wing	4	60	115
Myopia	1	13	17
	2	13	49
	3	13	77
	4	13	106
—index	2	17	50
—pathological axial	1	17	18
	3	17	78
Neutralisation	1	9	16
	2	9	48
Squints	1	55	25
—Hirschberg test	1	41	22
	2	41	55
Transposition	1	26	18
	2	21	51
Use of prisms	2	37	54
	1	37	22
Visual acuity	1	24	19
	2	24	51
	2	26	52
	3	24	79
	4	24	108
	4	40	111
Worth's four-dot test	3	41	83
	4	41	111
Spectacle prescription			
Achromats	4	19	107
Ametropia	3	21	78
Aphakia	1	22	19
Back vertex distance	1	23	19
and power calculations	1	11	17
	2	11	49
	2	58	58
	3	11	77
	4	11	106
	4	23	108
Back vertex power	2	23	51
Best formed lenses	1	59	26
	2	59	58
Decentration	1	54	25
	2	54	57
	3	54	85
Different ages	3	18	78
Interpupillary distance	2	57	58

Topic	Test	No.	Page No.
Intra-ocular lens			
advantages	4	22	108
contra-indications	4	58	115
Isekonic lenses	1	43	23
	2	43	55
Materials used	1	57	25
Myopia	4	54	114
Presbyopia	4	18	107
	3	19	78
	1	42	23
	2	42	55
	3	42	83
—Varilux lens	4	42	112
Prisms	4	55	114
Prisms—Fresnel	1	20	18
Relative spectacle			
magnification	1	52	24
	2	52	57
Ring scotoma	1	51	24
Tinted spectacles	3	57	86
Optical aberrations			
General	1	27	20
Chromatic	3	27	80
Image distortion	2	27	52
Oblique astigmatism	4	37	109
Spherical aberrations	4	27	111
	2	51	57
	3	52	85
Trial frames	4	43	112
Trial lens	3	43	83
Contact lens			
General	1	46	23
	2	46	56
	2	50	56
	3	50	84
	4	49	113
Advantages	3	47	84
Anisometropia	3	60	86
Aphakia	2	22	51
Astigmatism	4	50	113
	3	51	85
	3	46	84
	4	46	112
Bifocal	1	48	24
Complications			
	2	48	56
—*Pseudomonas*	3	58	86
Extended wear	3	49	84

Topic	Test	No.	Page No.
Keratoconus	4	47	112
Materials	1	50	24
Soft	2	49	56
	1	47	24
	2	47	56
Stabilisation	4	57	114
Refraction			
General	1	30	20
Pinhole	1	29	20
Objective			
	1	31	20
	2	31	53
	2	28	52
	3	28	80
	4	28	109
	3	26	79
	3	31	80
	3	38	82
	4	32	110
large refractive error	3	32	81
opacities	2	32	53
scissors shadow	1	33	15
Subjective			
general	4	26	109
	1	28	20
accommodation reduction	3	29	80
	3	30	80
astigmatism—axis	3	15	77
	4	30	109
astigmatic dial	2	33	21
actigmatic fan	3	33	81
binocular test	2	56	58
cross-cylinder	1	25	19
	4	25	108
Duochrome test	1	26	19
	3	25	79
fogging techniques	1	32	21
stenopaeic slit	2	25	52
Optical instruments			
General	4	45	108
CCTV	1	45	23
Focimeter	1	38	22
	2	38	54
Contact lens checking	1	49	24
Geneva lens measure	4	38	111
Hruby lens	3	39	82

Topic	Test	No.	Page No.
Keratometry	1	60	26
Magnifier	2	45	56
	3	45	83
Pachymeter	2	39	54
Ophthalmoscopy			
general	1	35	21
direct	2	34	53
	1	34	21
	4	34	110
indirect	3	34	81
	2	35	54
	3	35	81
	4	35	110
	4	48	113
Placido's disc	4	29	109
Purkinje's image	4	59	115
Retinoscope-			
streak	2	30	52
Slit lamp	3	40	82
	2	40	55
Telescope			
—astronomical	1	44	23
	2	44	55
—Galilean	3	44	83
	4	44	112
Three-mirror contact lens	4	39	111
Tonometer	1	40	22